SUSTAINABLE ENERGY

SUSTAINABLE ENERGY
Towards a Zero-Carbon Economy using Chemistry, Electrochemistry and Catalysis

JULIAN R.H. ROSS
Emeritus Professor, University of Limerick, Limerick, Ireland; Member of the Royal Irish Academy (MRIA); Fellow of the Royal Society of Chemistry (FRSC)

ELSEVIER

Elsevier
Radarweg 29, PO Box 211, 1000 AE Amsterdam, Netherlands
The Boulevard, Langford Lane, Kidlington, Oxford OX5 1GB, United Kingdom
50 Hampshire Street, 5th Floor, Cambridge, MA 02139, United States

Copyright © 2022 Elsevier B.V. All rights reserved.

No part of this publication may be reproduced or transmitted in any form or by any means, electronic or mechanical, including photocopying, recording, or any information storage and retrieval system, without permission in writing from the publisher. Details on how to seek permission, further information about the Publisher's permissions policies and our arrangements with organizations such as the Copyright Clearance Center and the Copyright Licensing Agency, can be found at our website: www.elsevier.com/permissions.

This book and the individual contributions contained in it are protected under copyright by the Publisher (other than as may be noted herein).

Notices
Knowledge and best practice in this field are constantly changing. As new research and experience broaden our understanding, changes in research methods, professional practices, or medical treatment may become necessary.

Practitioners and researchers must always rely on their own experience and knowledge in evaluating and using any information, methods, compounds, or experiments described herein. In using such information or methods they should be mindful of their own safety and the safety of others, including parties for whom they have a professional responsibility.

To the fullest extent of the law, neither the Publisher nor the authors, contributors, or editors, assume any liability for any injury and/or damage to persons or property as a matter of products liability, negligence or otherwise, or from any use or operation of any methods, products, instructions, or ideas contained in the material herein.

Library of Congress Cataloging-in-Publication Data
A catalog record for this book is available from the Library of Congress

British Library Cataloguing-in-Publication Data
A catalogue record for this book is available from the British Library

ISBN: 978-0-12-823375-7

For information on all Elsevier publications
visit our website at https://www.elsevier.com/books-and-journals

Publisher: Susan Dennis
Acquisitions Editor: Anita Koch
Editorial Project Manager: Charlotte Kent
Production Project Manager: Bharatwaj Varatharajan
Cover Designer: Christian J. Bilbow

Typeset by STRAIVE, India

Contents

Preface	*vii*
Acknowledgements	*ix*

1. Introduction — 1

Energy production and the greenhouse effect	1
Greenhouse gases	4
Consequences of the greenhouse effect	9
The sources of greenhouse gas emissions	11

2. Traditional methods of producing, transmitting and using energy — 21

Introduction	21
Coal	21
Crude oil	38
Natural gas	42
Concluding remarks	47

3. Less conventional energy sources — 49

Introduction	49
Nuclear energy	50
Geothermal energy	55
Tidal energy	59
Wave power	62
Hydroelectric power	63
Wind power	67
Solar power	69
Concluding remarks	75

4. The production and uses of hydrogen — 77

Introduction	77
The production of hydrogen from natural gas by steam reforming	77
The production of hydrogen from natural gas by other methods	89
Methanol production	96
Production of fuels using the Fischer Tropsch process	97
Production of ammonia	99
Conclusions	102

v

vi   Contents

5. Biomass as a source of energy and chemicals   **103**

Introduction   103
Wood as a source of energy and paper   104
Non-traditional uses of biomass: First and second generation bio-refinery
     processes   112
Concluding remarks   129

6. Transport   **131**

Introduction   131
Historical development of mechanically driven transport   131
Exhaust emission control   143
Hybrid vehicles   150
Plug-in hybrid vehicles   152
Battery electrical vehicles   154
Fuel cell vehicles   159
Concluding remarks   160

7. Batteries, fuel cells and electrolysis   **163**

Introduction   163
The Volta pile, Faraday and the electrochemical series   163
Half-cell EMF's and the electrochemical series   167
The kinetics of electrochemical processes   170
Electrochemical batteries   175
Flow batteries   185
Fuel cells   186
Electrolysis   192

8. The way forward: Net Zero   **197**

Introduction   197
Hydrogen production using renewable energy   199
Fuel cells to be used for transportation purposes   203
Solid oxide hydrolysis cells (SOEC's) for hydrogen production and their use
     for the synthesis of green ammonia and methanol   205

Tailpiece   *221*
Index   *225*

Preface

It is not possible to open a newspaper or magazine without reading of some aspect of the global problem of climate change and of the measures that are necessary to combat it so that we can achieve 'zero carbon' before the year 2050. There has been a steady increase in the emission of greenhouse gases since the Industrial Revolution and the aim of all those countries that have signed up to the Paris Accord is to bring back the resultant temperature rise to no more than 2°C (and even to 1.5°C) within fewer than 30 years.

This book considers many aspects of the potential uses of 'sustainable energy'. In this context, this is the energy that can be obtained by using renewable resources such as wind power, hydroelectric power or solar radiation, and the book discusses how this energy can be used in place of conventionally derived energy from fossil reserves: coal, oil and natural gas. In order to set the scene, the book also discusses in some detail the many ways in which conventional energy is currently used.

The first chapter sets the scene by considering some aspects of the greenhouse effect and outlines the objectives of the Paris Accord that is aimed at reducing the emissions responsible for the effect. It then traces the origins of the greenhouse effect, discussing some human activities (many of which are discussed later in the book) that have taken place since the Industrial Revolution and have contributed to the increased emissions.

The book then considers some important existing industrial activities, all related to the use of energy created from the use of fossil fuels, coal oil and natural gas, each of which results in the emission of greenhouse gases. Some of these emissions can be reduced by methods such as carbon collection and storage, but an alternative is to produce some of the chemicals and fuels on which we rely by using biomass-derived materials. Hence, the use of biomass as a source of energy and chemicals is then considered.

Transport, in one form or the other, is responsible for a significant share of our greenhouse gas emissions. The developments that have occurred since the Industrial Revolution of various forms of transport are outlined and modern developments such as the use of hybrid engines, battery power and fuel cells are then considered. This leads to a detailed discussion of various types of batteries and fuel cells followed by a section considering the potential importance of electrolysis brought about using renewable energy as a means of producing hydrogen and syngas.

vii

The final chapter considers how green hydrogen or syngas produced using electrolytic methods fuelled by renewable electricity can be used in industrial applications such as ammonia and methanol synthesis, the production of steel and cement manufacture. It also considers the importance of achieving reductions in emissions from commercial, domestic and agricultural sources.

The reductions required to allow us reach the targets set in the Paris Accord are enormous and the progress towards achieving these aims has been disappointingly slow until now. Governments and responsible agencies must therefore pay significantly greater attention to ways in which objectives can be achieved and can only manage that by applying the 'carrot and stick approach': offering incentives to all energy users that encourage energy-saving initiatives and the introduction of new methods while at the same time penalising inactivity.

Acknowledgements

As I did in my previous two books, I first thank the very many people with whom I have worked over the years for their efforts and enthusiasm, especially the students and postdocs from my various research groups, too many to name individually, who have helped me build up my knowledge of catalysis and related fields. Thanks are also due to the many scientists and engineers with whom my different research groups have collaborated and from whom I have learnt much about the applications and exploitation of fundamental research in the field of heterogeneous catalysis. This collaborative work was carried out with funding provided by many sources, particularly by various EU research programmes.

I thank Elsevier and the many people from that company with whom I have collaborated during my editorial work for *Applied Catalysis* and *Catalysis Today* and in the production of the three books that I have now written and published with them. In particular, I thank Kostas Marinakis who not only guided me through the process involved in the planning of this book but with whom I have had many previous interactions during my work as an editor. I wish him well in his retirement. Thanks are also due to Kostas's successor, Anita Koch, for her more recent involvement with the production of this book; to Narmatha Mohan for her assistance in ensuring that the necessary permission had been obtained to reproduce copyright material; and to Bharatwaj Varatharajan for his careful and helpful work on the final production and during the proofreading stage. I particularly thank Alice Grant who, as the most recent Elsevier desk editor involved, has cheerfully and helpfully worked with me for most of the writing process.

My thanks are due to two good friends who, each in particular way, helped me during the writing phase: firstly, my colleague and long-standing collaborator, Michael Hayes, who very kindly read through the first draft of Chapter 5 (Biomass as a Source of Energy and Chemicals) and not only provided me with useful comments but also gave me invaluable information on soil organic matter; and secondly, Tony Hilley, a retired offshore oil and gas engineer, who encouraged me throughout the writing phase by providing me with a large number of important web links to recent developments in the field of energy. I also thank Miguel Bañares for his comments on the contents of the completed manuscript and for suggesting the term 'Mount Sustainable'.

x   Acknowledgements

Finally, I must once more express my sincere thanks to my wife, Anne, who has encouraged and supported me during the writing of yet another book. This support was even more important for the current volume as she has patiently tolerated my involvement in the task during a period when COVID-19 intruded on our existence and forced long periods of self-isolation.

Julian R.H. Ross

CHAPTER 1

Introduction

Energy production and the greenhouse effect
Solar activity and global warming

For centuries, we have relied on our natural resources for the provision of energy. Early man relied on the combustion of biomass (predominantly wood) to provide heat and fuel for cooking. Very much later, roughly at the time of the Industrial Revolution, he discovered coal, oil and natural gas and these discoveries led to our current almost total dependence on fossil fuels for the provision of energy.[a] Until the Industrial Revolution, the earth's population was predominantly agrarian and any fluctuations in climate that occurred were related only to variations in solar activity. Since then, however, there has been a steady increase in the average global temperature and it is now generally recognised that this change of temperature is related to increased emissions of the so-called greenhouse gases.

Fig. 1.1 shows the values of the solar irradiance and also the global temperature that have been measured over the period since 1880; although there have been some significant changes in the solar activity (and there was a marked maximum value around 1960), the measured values have remained relatively steady over the last 50 years. However, there has been a very significant increase in global temperature during the same period. It is now generally accepted (see Fig. 1.2) that human activities have been responsible for this increase in temperature.[b]

---

[a] We also rely on petroleum derivatives for the manufacture of many of the other resources that we now take for granted: polymers, dyestuffs, pharmaceuticals, detergents, etc. However, our fossil fuel reserves are gradually diminishing and they must therefore be used much more strategically.

[b] A useful summary of some aspects of climate change are to be found in the publication "Vital Climate Change Graphics" published by UNEP/GRID-Arendal; this is available as a free pdf from https://www.grida.no/publications/254/.

Sustainable Energy
https://doi.org/10.1016/B978-0-12-823375-7.00006-8

Copyright © 2022 Elsevier B.V.
All rights reserved.

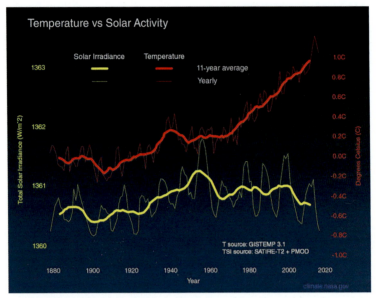

Fig. 1.1 Global temperature and solar activity since 1880. The yearly variations of both these parameters are shown by lighter curves and these have been averaged to give the more distinct curves. *(Source: https://climate.nasa.gov/.)*

Fig. 1.2 IPPC key findings. Predicted major changes due to global warming. *(Source: https://climate.nasa.gov/.)*

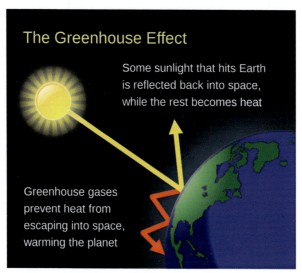

Fig. 1.3 Schematic representation of the greenhouse effect. *(Source: From Wikipedia (https://en.wikipedia.org/wiki/Greenhouse_effect/).)*

The greenhouse effect

Much life on earth as we know it depends on the light radiation from the sun that penetrates through the atmosphere to warm the earth's surface. Without the atmosphere, much of the incident radiation would be re-emitted from the surface and would be totally lost in space. Fortunately however, the atmosphere acts in the same way as does the glass in a greenhouse,[c] absorbing and reflecting back much of the re-emitted radiation and ensuring that the temperature of the atmosphere is increased. This process is shown schematically in Fig. 1.3. The resultant temperature on earth is a delicate balance of the levels of incoming and reflected radiation and is thus very susceptible to changes in the composition of the atmosphere; if too much of the reflected radiation is retained by the atmosphere, the temperature of the earth will rise.

---

[c] With a greenhouse, almost all the incident light passes through the glass and is absorbed by the soil within the structure; some of the energy is then re-emitted at a different wavelength but this is now absorbed by the glass, ensuring that the increased temperature in the greenhouse is maintained.

4   Sustainable energy

Greenhouse gases

Table 1.1 lists the main greenhouse gases associated with global warming, giving for each the chemical formula, the global warming potential relative to that for CO_2 over a 100-year lifespan and the atmospheric lifetime in years.

Table 1.2 shows the main sources of these greenhouse gases and also gives the pre-industrial atmospheric concentrations and the current atmospheric concentrations. The first three gases all existed in the pre-industrial era, although the concentrations have all increased since, while the last entries all refer to man-made gases introduced over the last century. We obtain an approximation to the relative contributions of the relevant gases to global warming if we multiply the current concentrations of each gas by the global warming potential from Table 1.1. The resultant figures show that the main culprits are CO_2, methane and nitrous oxide: not taking into account the small contributions of the fluorine-containing molecules, CO_2 contributes 73.3% of the total global warming potential of these gases while methane contributes 8.5% and N_2O contributes 18.2%. Although the contributions of the various fluorinated molecules are relatively low, it needs to be recognised that the lifetimes of these species are significantly above those of the other greenhouse gases and it is for this reason that they are no longer manufactured. As we will see below, there are a number of other greenhouse gases, some of which contribute to global warming while others do not. Water vapour is one example of a gas which does not contribute directly to global warming and ozone is

Table 1.1 Global warming potential and atmospheric lifetime for the most important greenhouse gases.

Greenhouse gas	Chemical formula	Global warming potential, 100-year time-span	Atmospheric lifetime/years
Carbon dioxide	CO_2	1	100
Methane	CH_4	25	12
Nitrous oxide	N_2O	298	114
Chlorofluorocarbon-12 (CFC-12)	CCl_2F_2	10,900	100
Hydofluorocarbon-23 (HFC-23)	CHF_3	1,48,800	270
Sulfur hexafluoride	SF_6	22,800	3200
Nitrogen trifluoride	NF_3	17,200	740

Reproduced from the Fourth Assessment Report (Intergovernmental Panel on Climate Change, IPCC, 2007).

Introduction 5

Table 1.2 The most important sources of the major greenhouse gases and their preindustrial and recent (2011) concentrations.

Greenhouse gas	Major sources	Pre-industrial concentration/ ppb	2011 concentration/ ppb
Carbon dioxide	Fossil fuel combustion Deforestation Cement production	278,000	390,000
Methane	Fossil fuel production Agriculture Landfills	722	1803
Nitrous oxide	Fertilizer application Fossil fuel and biomass combustion Industrial processes	271	324
Chlorofluorocarbon-12 (CFC-12)	Refrigerants	0	0.0527
Hydofluorocarbon-23 (HFC-23)	Refrigerants	0	0.024
Sulfur hexafluoride	Electricity transmission	0	0.0073
Nitrogen trifluoride	Semiconductor manufacturing	0	0.00086

another. We will now consider each greenhouse gas in turn, starting with water vapour.

Water vapour

The most important greenhouse gases are water vapour and carbon dioxide. Both of these result from the combustion of fossil fuels but may also arise from other sources. Water-vapour, which results predominantly from the evaporation of surface water, has a feedback effect: it forms clouds in the atmosphere and these lead to precipitation, this having the consequence that the level of water-vapour in the atmosphere is well controlled. The clouds also reflect some of the radiation (UV, visible and infra-red) reaching the

6   Sustainable energy

atmosphere from the sun, this also restricting the temperature rise. One consequence of the presence of increased partial pressures of carbon dioxide in the atmosphere (see below) is that the resulting temperature rise also causes an increase in the partial pressure of the water in the atmosphere, thus giving rise to a further increase in the temperature. Hence, water vapour has an indirect effect on global warming.

Carbon dioxide

Even though the concentration of carbon dioxide in the atmosphere is much lower than that of water, its effect is much greater since there is no equivalent feedback mechanism to that with water: once the carbon dioxide reaches the atmosphere, its residence time there is very much greater than that of water. The double bonds of the $C{=}O$ linkages of the CO_2 absorb much of the infrared radiation emitted from the earth and prevent this radiation from leaving the atmosphere. The result is an increase in atmospheric temperature. It should be recognised that the CO_2 reaching the atmosphere can come from many sources apart from combustion, for example, respiration and volcanic eruptions. It can also arise from deforestation and changes in land use. As discussed above, the increase in atmospheric temperature caused by the CO_2 also has an effect on the level of water vapour in the atmosphere since the saturation vapour pressure of the water increases with increasing temperature and hence this magnifies the effect of the increase in CO_2 concentration. Atmospheric CO_2 is essential for the growth of plants and all types of vegetation. Hence, we rely on a steady partial pressure of CO_2 to enable agricultural activities. We will return to the subject of CO_2 utilisation in subsequent chapters. As shown in Fig. 1.4 of Box 1.1, there has been a dramatic increase in the concentration of CO_2 in the atmosphere over the last 70 years.

Methane

Methane (CH_4), the simplest hydrocarbon molecule, may arise from a number of sources, both natural and man-made. It is produced by the decomposition of wastes in landfills, from agricultural sources such as rice paddies, from digestive processes of ruminants (e.g. cattle and sheep) and manure management from domestic livestock. It was also commonly emitted as waste from oil well operations and as leakages from chemical processing; however, both of these sources are now much more carefully controlled. (See Box 1.2 for an example of methane emission.) Methane is a much more active greenhouse gas than is CO_2, its 'global warming potential' being much higher (see Table 1.1). The atmosphere also contains yet lower concentrations of other hydrocarbons such as the vapours of petroleum and diesel components and these too are greenhouse gases. Methane and the other hydrocarbons have much longer lifetimes than does CO_2 in the atmosphere; while CO_2 is removed by natural processes, the

BOX 1.1 Variation of global CO_2 concentrations as a function of time

Fig. 1.4 shows the concentration of CO_2 in the atmosphere as a function of time over many centuries. These data have been compiled from the analysis of air bubbles trapped in ice over the last 400,000 years. During ice ages, the levels were about 200 ppm (ppm) and they rose to around 280 ppm in the warmer interglacial periods. The rise after about 1950 is attributable to a rapid increase in the use of fossil fuels as will be discussed further in later sections.

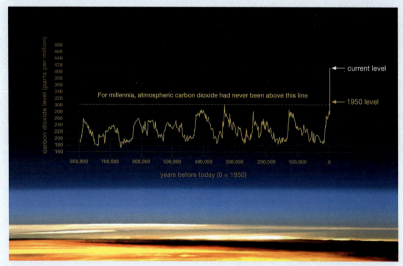

Fig. 1.4 The variation in carbon dioxide concentration as a function of time. It is clear that there has been a dramatic increase in the level of carbon dioxide since 1950 that is well outside the normal temporal variations. *(Source: https://climate.nasa.gov/resources/)*

8   Sustainable energy

> ### BOX 1.2 Methane emissions from the production of bitumen from oil sands
>
> There is a significant industry based on the extraction of bitumen from underground reservoirs containing oil sands. The bitumen is heated using the injection of steam to decrease its viscosity and to make it flow more easily. The steam is generated by the combustion of natural gas (methane) and this process gives rise to significant CO_2 emissions, these contributing to global warming. Canada's Oil Sands Initiative Alliance (COSIA) is attempting to find ways of reducing these emissions and has announced that it will assist innovators in developing new routes to reduce the emissions formed during the steam generation step, preferably producing a sequestration-ready product (e.g. concentrated CO_2) or a saleable product (e.g. carbon black).
>
> https://cosia.ca/blog/helping-clear-air-oil-sands-emissions-natural-gas-decarbonization/

hydrocarbons are relatively stable. As a result, they all have higher global warming potentials (Table 1.1).[d]

Nitrous oxide

Nitrous oxide (N_2O) is a very powerful greenhouse gas (see Table 1.1) that is formed by soil cultivation practices, especially by the use of nitrogenous fertilisers; it is also formed by fossil fuel combustion, nitric acid production and biomass burning. It should be recognised that N_2O is only a minor constituent of so-called NO_x, a mixture of the oxides of nitrogen (N_2O, NO and NO_2), formed in high-temperature combustion processes[e] such as those involved in electricity generation and internal combustion engines. NO_x is considered to be an atmospheric pollutant and its emission is associated with the formation of 'acid rain'; the NO_x emissions from these sources are generally controlled by catalytic reduction processes.[f]

---

[d] It should be noted that the global warming potential of methane is time dependent as it is gradually destroyed by oxidation processes in the atmosphere; over periods less than 100 years, the value of the global warming potential is much larger.

[e] At high temperatures and in excess oxygen, thermodynamics favours the formation of NO_2.

[f] For a description of the control of NOx emissions from power stations and automobiles, see Contemporary Catalysis – Fundamentals and Current Applications, Julian R.H. Ross, www.elsevier.com/books/contemporary-catalysis/ross/978-0-444-634740-0. See also Chapter 2.

Ozone and chlorofluorocarbons

Ozone is also a greenhouse gas. It is formed in the troposphere by the interaction of sunlight with other emissions such as carbon monoxide or methane and also by interaction with hydrocarbons and NOx from automobile emissions. The lifetime of ozone is relatively very short (days to weeks) and its distribution is very variable. It absorbs harmful UV radiation and we are therefore dependent on the presence of the ozone layer. The creation of an ozone hole over the Antarctic is ascribed to the emission of chlorofluorocarbons (CFCs), another class of powerful greenhouse gas, and this has led to the banning of the production of these molecules; the production of other hydrofluorocarbons (HFCs) and perfluorocarbons (PFCs) is also being phased out.[g]

Consequences of the greenhouse effect

It is generally recognised that it is difficult to predict the consequences of changing the composition of the naturally occurring atmospheric greenhouse that surrounds the Earth. However, it is extremely likely that the average temperature of the Earth will continue to rise; even though some areas will become cooler, others will become warmer. Warmer conditions will probably give rise to more evaporation and precipitation although some regions will become wetter and others will become drier. A stronger greenhouse effect will warm the world's oceans and these will expand and increase sea levels; additionally, glaciers and other ice will melt, thus further increasing the sea level. The increased CO_2 concentration in the atmosphere will encourage some crops and other plants to grow more rapidly and to use water more efficiently; however, at the same time, higher temperatures and change in the climate patterns may cause changes in the distribution of the areas where crops grow best. Although climate change has been the subject of great concern for quite some time, it was only about 30 years ago that scientists became particularly concerned about the changes which were occurring;[h] see Box 1.3. Arrhenius discussed in 1886 the importance of the increases in emissions of carbon dioxide resulting from coal-burning; he argued that this would lead to improved agricultural practices and better

---

[g] This phasing out is part of the Kyoto Protocol (2005); the US has not ratified this international agreement.

[h] An excellent article by Andrew Revkin outlining some of the history of awareness of the problems of climate change is to be found in the National Geographic Magazine of July 2018 (https://www. national geographic.com/magazine/2018/07/embark-essay-climate-change-pollution-revkin/). However, there are many other such articles available on the web.

10   Sustainable energy

growth of crops. An article by Waldemar Kaempffert in the New York Times as early as 1956 (October 28)[i] predicted that the increased emissions from energy production would lead to long-lasting environmental changes. This article pointed out very clearly that an impediment to counteracting these changes was the abundance of coal and oil in many parts of the world and that these fossil fuels would continue to feature in industrial use as long as it was financially beneficial to use them. However, it was not until 1988 that the World Meteorological Organisation (WMO) established the Intergovernmental Panel on Climate Change (IPCC). The IPCC summarises the scientific developments in countering climate change in the IPCC Assessment Reports that are published every five to six years, these being compiled in association with a number of other related reports from the panel. (The Sixth Synthesis Report is due in 2022.)[j] In parallel to these activities, there have been a large number of reports by other agencies, both international and national, some of which will be quoted below. There have also been two important international agreements on how climate change should be counteracted, the Kyoto Agreement of 1992 and the Paris Agreement of 2015, both established under the auspices of the United Nations Framework Convention on Climate Change (UNFCCC). Just fewer than 200 countries were signatories to these agreements, the aims of which being to reduce the emission of greenhouse gases.[k] These agreements will be discussed further below.

BOX 1.3 The importance of the existence of a greenhouse effect on earth

If the earth did not have an effective shielding greenhouse layer containing high levels of water vapour as well as CO_2 and the other greenhouse gases, much of the light of all wavelengths reaching the surface of the earth from the sun would be reflected back into space without warming the planet. The planet Mars has a relatively thin atmosphere consisting largely of carbon dioxide but little or no water vapour and this results in a weak greenhouse effect. As a
Continued

---

[i] See New York Times December 8, 2015: https://www.nytimes.com/interactive/projects/cp/climate/2015-paris-climate-talks/from-the-archives-1956-the-rising-threat-of-carbon-dioxide.

[j] The Montreal Protocol to reduce the emission of compounds that affect the ozone layer such as the chlorofluorocarbons mentioned above was agreed in 1987.

[k] The US under President Trump had announced that it was going to leave the Paris Agreement but that decision has now been reversed by President Biden; of the countries with over 1% share of the global emissions, only Iran and Turkey are not parties to the agreement.

> ### BOX 1.3 The importance of the existence of a greenhouse effect on earth—cont'd
>
> result, the surface of Mars is largely frozen and there is no evidence of any life form. In contrast, Venus has a much higher concentration of CO_2 in its atmosphere (150,000 times as much as on Earth and 19,000 as much as on Mars) and the surface temperature is +460°C. Again, this atmosphere would not be amenable to life as we know it. (See https://agreenerfutureblog.wordpress.com/1-the-natural-greenhouse-effect/1-4-greenhouse-effect-on-other-planets/; https://earthsky.org/space/venus-mars-atmosphere-teach-us-about-earth/)

The global emission of all greenhouse gases in 2010 was 48 gigatonnes of CO_2 equivalent and it was estimated that this figure would increase to 53.5 gigatonnes in 2020. What is much more alarming is that the amount will increase to 70 gigatonnes by 2050 unless action is taken to reduce greenhouse emissions. These unfettered increases will give rise to totally unacceptable global temperature increases, these in turn giving rise to myriad problems for the world's inhabitants. The Kyoto Protocol and the Paris Agreement have led to two targets relating to temperature rise: the first, a limit of 2.0°C compared to the emissions in the pre-industrial era; and the second, pursuing at the same time means to limit the temperature increase to 1.5°C.

Fig. 1.5 illustrates the predicted changes occurring for different scenarios envisaged in the run-up to the Paris Accord, ranging from the absence of any policy (resulting in a very high chance that the global temperature rise will be well above 4°C) to a very strict policy (when the chance of the temperature approaching pre-industrial levels will be much greater). The solid curves of the diagram show the predicted emissions of CO_2 for different scenarios, from no action (top curve) to the most ambitious series of actions with higher rates of decarbonisation (1.5°C warming, bottom curve). It will be seen that only the latter approach gives any significant reduction in the emission of greenhouse gases. The slightly lower set of ambitions, with constant rates of decarbonisation (2°C warming) gives a levelling off of the levels of CO_2 after about 2030.

The sources of greenhouse gas emissions

There have been a number of very detailed national and international reports that give information on emissions of greenhouse gases and list the main

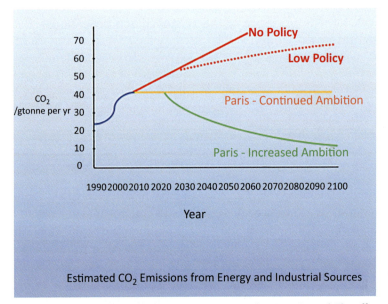

Fig. 1.5 Changes in carbon dioxide emissions envisioned in Paris Accord. The effects on CO_2 emissions of either no change or adopting various strategies (see text). *(Adapted from Climate Science Special Report (US Global Change Program), Fourth National Climate Assessment (NCA4) Vol. 1, Chapter 14.2: https://science2017.globalchange.gov/.)*

sources. Some of these reports will be discussed in more detail in later sections. One of the most relevant of these reports for the present purposes is one by the European Environment Agency[1] and the following sections will summarise some of its most relevant material.

Table 1.3 shows data for the total emissions of greenhouse gases from European countries in 2017 and also shows the changes in emissions that have occurred in the period between 1990 and 2017, the greenhouse gas emissions per capita and the change in the total energy intensity of each country in the period 1990–2017. The majority of these countries are members of the EU but the data for Norway and Turkey have also been added for completeness. It should be recognised that the United Kingdom is also included as an EU country as the data relate to a period prior to BREXIT. (It should be noted that the data for a number of smaller EU countries have been omitted for clarity; full details are available in the source report which also gives some more detail for each country.) The figures for the whole EU are given in the last row; the EU figures for the % changes in GHG

---

[1] The European Environment – State and Outlook 2020, European Environment Agency, (2019), doi: https://doi.org/10.2800/96749, downloadable from https://eea.europa.eu/.

Table 1.3 Greenhouse gas emissions from EU countries.

Country	Total GHG emissions 2017/ MtCO$_2$ equiv.	Change in GHG emissions 1990–2017/%	GHG emissions per capita in 2017/ tCO$_2$ equiv. per person	Change in the total energy intensity of the economy 1990–2017/%
Austria	84.5	+6.2	9.6	−18.3
Belgium	119.4	−20.3	10.5	−27.1
Bulgaria	62.1	−39.5	8.8	−54.0
Czech Republic	130.5	−34.7	12.3	−48.4
Denmark	50.8	−29.5	8.8	−35.5
Finland	57.5	−20.5	10.4	−24.5
France	482.0	−13.4	7.2	−25.5
Germany	936.0	−25.9	11.3	−40.1
Greece	98.9	−6.4	9.2	−13.0
Hungary	64.5	−31.5	6.6	−38.5
Ireland	63.8	+12.9	13.3	−66.1
Italy	439.0	−15.9	7.3	−10.8
Netherlands	205.8	−9.1	12.0	−34.2
Poland	416.3	−12.4	11.0	−61.7
Portugal	74.6	+22.8	7.2	−4.0
Romania	114.8	−53.9	5.9	−69.6
Slovakia	43.5	−40.8	8.0	−63.6
Spain	357.3	+21.8	7.7	−14.3
Sweden	55.5	−23.7	5.5	−39.8
UK	505.4	−37.6	7.7	−49.3
(Norway	*54.4*	*+4.9*	*10.3*	*−22.4)*
(Turkey	*537.4*	*+144.5*	*6.7*	*−12.8)*
EU-28	**4483.1**	**−21.7**	**8.8**	**−36.3**

The data for two non-EU countries, Norway and Turkey, are also included in italics and the total emissions from the EU (EU-28) are shown in bold figures.
Some data concerning greenhouse gas emissions from some European countries.

emissions, the GHG emissions per capita and the change in the energy intensity of the economy allow one to see how each country has performed compared with the average result for the whole EU. It can be seen that Bulgaria, the Czech Republic, Germany, Ireland, Poland, Romania, Slovakia and the UK have performed better than the average.

Fig. 1.6 shows the emissions of the principal greenhouse gases per source type (given as CO_2 equivalents in millions of tons per year) for all the EU states for the period 1990 to 2017 and Table 1.3 gives some additional information. Most of the categories included in the figure and table have shown

14   Sustainable energy

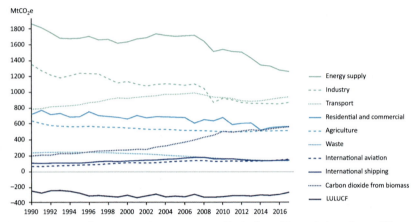

Fig. 1.6 EU greenhouse emissions per sector. Equivalent CO_2 emissions from different European sectors over the period 1990 to 2017. *(Source: Europe environment state and Outlook 2020 (https://www.eea.europa.eu).)*

significant decreases in emissions. A particularly large decrease occurred in energy supply, this being the result of increased use of renewable sources of energy, particularly wind and solar power.[m] The introduction of these technologies has been accompanied by a decrease in the use of coal combustion for electricity production. There has also been a significant drop in industrial energy consumption, due largely to improved efficiencies in industrial processes and to a switch away from the use of coal and oil to natural gas for the supply of energy, this change also being associated with improvements of technologies (Chapter 3). Some EU countries have been at the forefront in decreasing greenhouse gas emissions. For example, as can be seen from Table 1.3, Germany, originally one of the greatest emitters of greenhouse gases in Europe, had reduced its emissions by 25.9% over the period to a level of 936 million tons of CO_2 equivalent. (This has been achieved partly by a significant decrease in the use of lignite as a fuel.) Unfortunately, however, some countries had increased their contributions; for example, Ireland's contribution increased by 12.9% to 63.8 million tons of CO_2 equivalent and that of Cyprus increased by 55.7% to 10.0 million tons of CO_2 equivalent. These rather different results appear to be related to differences in the structures of the different economies. For example, Ireland has very little heavy industry but has a strong agricultural economy as well as

---

[m] Nuclear energy is also considered to be a renewable source of energy. France is particularly reliant on nuclear power. Germany, on the other hand, has decided to close down all of its nuclear reactors and is well on its way to doing so. Other countries such as Ireland have never had nuclear facilities.

large numbers of new high-technology companies consuming relatively large quantities of energy.

Table 1.4 gives a breakdown of the changes that have occurred in the emissions of greenhouse gases (both CO_2 and others) for the period 1990 to 2017 from all European countries, the data being given in million tons of CO_2 equivalents. It can be seen that road transportation continues to give increased emissions as do refrigeration and air conditioning. In all other sectors, there have been significant reductions; particularly important reductions have been found in residential heating, iron and steel production, the manufacturing industries and public electricity and heat production. Many of these improvements will be discussed in subsequent sections and chapters.

Fig. 1.6 shows that there has been a slightly decreasing contribution to greenhouse emissions from agriculture since 1990. These emissions, mainly from ruminants (cattle and sheep) are still significant and are therefore a continuing cause for concern, particularly in Ireland. A similar set of figures will apply globally although there will be national differences arising from different

Table 1.4 EU emissions per source.

Emission source	MtCO$_2$ e
Road transportation	170
Refrigeration and air conditioning	93
Aluminium production	−21
Agricultural soils: direct emissions of N_2O from managed soils	−22
Cement production	−26
Fluorochemical production	−29
Fugitive production from natural gas	−37
Commercial/institutional	−38
Enteric fermentation—cattle	−43
Nitric acid production	−46
Adipic acid production	−56
Manufacture of solid fuels and other energy industries	−60
Coal mining and handling	−66
Managed waste disposal sites	−73
Residential—fuels	−115
Iron and steel production	−116
Manufacturing industries	−253
Public electricity and heat production	−433
International aviation	*+89*
International navigation	*+35*

The emissions arising from international aviation and navigation outside the EU are also shown (in bold italics) for completeness.
Trends in EU emissions from the predominant sources in the period 1990 to 2017.

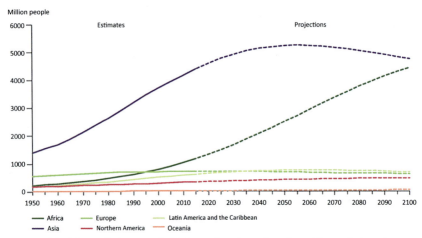

Fig. 1.7 World population trends. Estimated and predicted world population changes from 1950 to 2100. *(Source: Europe environment state and Outlook 2020 (https://www.eea.europa.eu).)*

methods of agriculture in each country. The global contribution to greenhouse gas emissions from agriculture is related to the very significant growth of population that has occurred over a couple of centuries since the start of the industrial revolution (Fig. 1.7).

Prior to the industrial revolution, the world population was only about 700 million. However, with improvements in food production and also in medical standards, the population grew rapidly to 1.6 billion people in 1900 and thereafter even more rapidly to reach 6 billion before the end of the twentieth century. The world population is already above 7 billion and it is projected to reach 8 billion by 2030. A significant proportion of the increased emission of greenhouse gases in the period since the industrial revolution has emanated from changes in land use that have been needed to provide food for the increasing population, particularly in relation to the use of fertilisers and the production of meat for human consumption. Hence, one aspect of the control of greenhouse gas emission in the future will have to be related to improvements in agricultural practice to reduce emissions such as those of methane from ruminants and N_2O from fertilizer applications. Of particular relevance to the present text is the conversion of biomass to valuable products: biofuels, industrially relevant chemicals and hydrogen. These topics will be considered in more detail in Chapter 5. The only category with negative emissions shown in Fig. 1.6 is LULUCF (land use, land-use change and forestry) in which CO_2 is consumed by the growth of

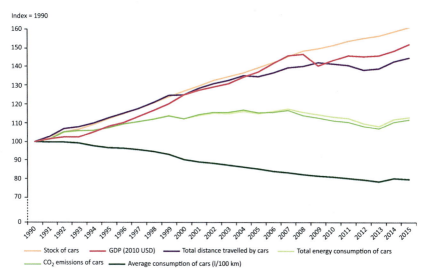

Fig. 1.8 European car usage. Changes (relative to data from 1990) in European car usage and performance over the period 1990 to 2017. *(Source: Europe environment state and Outlook 2020 (https://www.eea.europa.eu).)*

vegetation. It is clearly important that deforestation is very strictly controlled internationally and those significant efforts are made to increase the area of land devoted to forestry and other crops with much more efficient and careful use of fertilisers. The subject of the bio-economy is one which will receive little further attention in this text but the interested reader is referred to an open-access book edited by Lewandowski and published by Springer which gives excellent coverage of the topic.[n]

It is interesting to note that the emissions from transport shown in Fig. 1.6 increased steadily up until about 2006 but that they have decreased significantly since then. This is further illustrated by the data of Fig. 1.8 which shows the emissions arising from the use of private vehicles over the period 1990 to 2017. Although there has been a marked increase in the numbers of private cars and in the total distances travelled over this period (these figures being closely related to the parallel improvement in GDP), the total energy consumption and the emission of CO_2 have not increased in the same way; both of these parameters increased somewhat until about 2006 but then decreased significantly. This can be explained by the very significant improvement in engine efficiency and the consequent

---

[n] "Bioeconomy – Shaping the Transition to a Sustainable, Biobased Economy", Edited by I. Lewandowski, ISBN 978-3-319-68152-8, https://doi.org/10.1007/978-3-319-68152-8.

decrease in the average consumption per vehicle that occurred over the period considered.

International aviation and international shipping (see Fig. 1.6) have both shown steady increases and it is related to these figures (which are duplicated in statistics for non-EU nations) that the carbon footprint of international travel has recently attracted so much concern. This is highlighted further in Fig. 1.9 which compares the energy usage for different transport modes; only the usage by international travel continues to increase while the other modes of transport have shown significant decreases since about 2005.

The funding of research and development work related to energy supply in Europe has shown some significant changes over the last 40 years. Fig. 1.10 shows the trends in European spending over that period for various different technologies. Up until the mid-1980s, the predominant area of research that was funded related to nuclear energy while research in fossil fuels also attracted significant research effort. Hydrogen and fuel cells attracted some added interest around 1990 but all research activities decreased somewhat from 1990 until the early years of the new century. Since about 2005, there have been marked increases in research activities, particularly significant research expenditures having occurred in the areas of renewables and energy efficiency. The funding

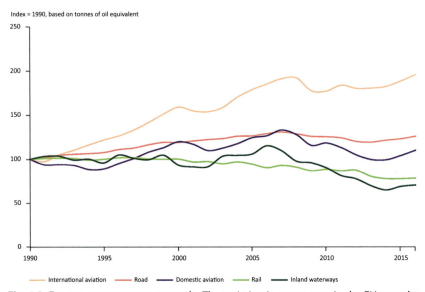

Fig. 1.9 Energy use per transport mode. The variation in energy use in the EU over the period 1990 to 2017 per transport mode relative to the usage in 1990. *(Source: Europe environment state and Outlook 2020 (https://www.eea.europa.eu).)*

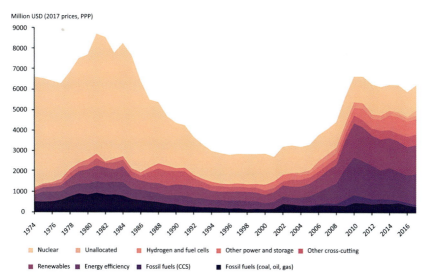

Fig. 1.10 European energy research funding. The funding of European energy research in various categories during the period 1974 to 2017. *(Source: Europe environment state and Outlook 2020. (https://www.eea.europa.eu/legal/copyright).)*

level for nuclear research (see Chapter 3), although much lower than in the period around 1980, is still significant. This is related to the fact that nuclear fusion has the potential, if ever achievable in a safe and controlled fashion, of supplying vast amounts of renewable energy without any undesirable by-products. Until fusion technology is available, however, we must seek other solutions to our energy requirements which will have minimal effect on our greenhouse gas emissions. We must also seek ways in which the emissions from traditional industries related to products such as iron and steel or cement can be reduced (Chapter 2).

This book, written from the point of view of a physical chemist, considers a number of important uses of energy, discussing in turn a number of chemical and other processes from which significant quantities of greenhouse gases are currently emitted. It then outlines strategies currently under development, or already in place, aimed at reducing these emissions. Particular emphasis is placed on the very considerable merits of a 'hydrogen economy' and hence special attention is given to the production, storage, transport and uses of hydrogen. Much of the material included is available in one form or another from the web and similar sources but often without an adequate discussion of the chemistry involved. Hence, one of the main aims of the book is to give a chemist's view of the technologies which are currently being explored and developed.

CHAPTER 2

Traditional methods of producing, transmitting and using energy

Introduction

The rapid growth of the highly industrialised world as we know it stemmed from the developments of new technologies introduced following the Industrial Revolution. Many of these technologies depended on the availability of sources of energy. In the early days of mechanisation associated with the Industrial Revolution, the major source of the energy required for the operation of the new machinery was coal, used to power steam engines; there was also some use of wind and water power. More recently, the world has come to depend more and more on other sources of energy, particularly oil and natural gas. Now, with the recognition of the problems of global warming, emphasis is being placed more and more on renewable energies: biomass, nuclear, solar, tidal, etc. As noted in Contemporary Catalysis—Fundamentals and Current Applications (2019), the use of these energy sources have followed a series of Kondratiev cycles (see Fig. 2.1).[a] Some aspects of each of these cycles will now be discussed briefly in turn.

Coal

All the sources the fossil fuels that we use derive originally from solar radiation since they were all formed in a series of geological processes from biomass, the growth of which depended on energy arriving from the sun. Coal was generated from layers of peat (formed in a process involving biomass degeneration) as a combustible sedimentary rock by the effect of pressure and heat over a period of millions of years. The world has very large reserves of coal. As shown in Table 2.1, the total global estimated reserves are of the order of 1055 billion tonnes and these are distributed reasonably equally

---

[a] It should be noted that the grouping labelled Nuclear, Solar, Tidal should also include other energy sources such as wind, hydroelectric and geothermal. The various sources of energy are discussed in more detail in this chapter.

Sustainable Energy
https://doi.org/10.1016/B978-0-12-823375-7.00007-X

Copyright © 2022 Elsevier B.V.
All rights reserved.

21

Kondratiev Cycles

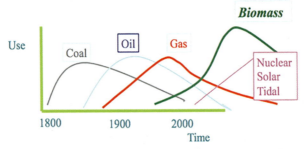

Fig. 2.1 Kondratiev cycles associated with the use of energy sources. Periodic cycles in the use of coal, oil, gas, biomass and renewable energy sources. *(Reproduced with the kind permission of Elsevier from Contemporary Catalysis—Fundamentals and Current Applications, J.R.H. Ross, 2018.)*

Table 2.1 Global coal reserves and expected lifetimes of these reserves.

Region	Total reserves/ million tonnes	Expected lifetime based on 2018 usage	Share of world total
North America	258,012	342	24.5%
US	250,219	365	23.7%
Central and South America	14,016	158	1.3%
Europe	134,593	215	12.8%
Germany	36,103	214	3.4%
Poland	26,479	216	2.5%
Ukraine	34,375	>500	3.3%
UK	29	11	<0.05%
CIS	188,853	329	17.9%
Kazakstan	25,605	217	2.4%
Russian Federation	160,364	364	15.2%
Middle East and Africa	14,420	53	1.4%
Asia Pacific	444,888	79	42.2%
Australia	147,435	304	14%
China	138,819	38	13.2
India	101,363	132	9.6
Indonesia	37,000	67	3.5
Total World	1,054,782	132	**100%**

Source: https://www.bp.com/content/dam/bp/business-sites/en/global/corporate/pdfs/energy-economics/statistical-review/bp-stats-review-2019-full-report.pdf.

throughout the world; as we will see later, the regions which have lower reserves (e.g. the Middle east and Africa) have larger reserves of oil and gas. It is interesting to note that the UK, where the Industrial Revolution commenced, has very low reserves. The known reserves of coal are made up of both 'hard coal' (anthracite and bituminous coal) and soft coal (sub-bituminous coal and lignite) and the breakdown of these for the major coal-producing countries are shown in Table 2.2.

As we will see, coal is used largely as a source of energy. The Kondratiev cycle of Fig. 2.1 associated with the use of coal is shown as having a maximum in the mid-1800s, coinciding roughly with its important contribution to the emergence of the Industrial Revolution. However, the use of coal is still very significant because of the large global reserves and hence it is still very important to the world economy; it is used very largely for energy generation although it can also be used as a source of chemicals.

Coal combustion for heating purposes

Coal is composed mainly of carbon but also contains some oxygen, hydrogen and nitrogen as well as sulphur. The combustion of coal therefore gives predominantly CO_2 from the reaction of the carbon content with oxygen of the air:

$$C + O_2 \rightarrow CO_2 \; \Delta H^\circ = -393.5 \, kJ/mol$$

Table 2.2 Hard and soft coal reserves of the major coal-producing countries.

Country	Hard coal/million tonnes	Soft coal/million tonnes	Total/million tonnes
USA	111,338 (23.3%)	135,305 (31.4)	246,643 (27%)
Russia	49,088 (10.3%)	107,922 (25.1%)	157,010 (17%)
China	62,200 (13%)	52,300 (12.2%)	114,500 (13%)
India	48,787 (10.2%)	45,660 (10.6%)	94,447 (10%)
Australia	38,600 (8.1%)	39,900 (9.3%)	78,500 (9%)
South Africa	48,750 (10.2%)	0 (0%)	48,750 (5%)
Ukraine	16,274 (3.4%)	17,879 (4.2%)	34,153 (4%)
Kazakhstan	28,151 (5.9%)	3128 (0.7%)	31,279 (3%)
Poland	14,000 (2.9%)	0 (0%)	14,000 (2%)

The table shows the reserves in million metric tonnes of hard coal (anthracite and bituminous coal) and soft coal (sub-bituminous coal and lignite) for each country; the percentages of total world reserves of each are shown in brackets.
Source: https://www.bp.com/content/dam/bp/business-sites/en/global/corporate/pdfs/energy-economics/statistical-review/bp-stats-review-2019-full-report.pdf.

In principle, the CO_2 produced can be trapped and stored (using carbon capture and storage, CCS, a topic to be discussed further in later chapters) but this technology is not yet sufficiently developed to permit it to be adopted for general use. Hence, the CO_2 emitted is one of the main sources of greenhouse gas (25% of global greenhouse gas emissions, see Preface). Another serious problem associated with the combustion of coal is that the nitrogen and sulphur contents give rise to the emission of NO_x (N_2O, NO and NO_2) and SO_2; these gases are generally accompanied by the formation of particulates and trace metals. All these emissions have serious effects on human health by contributing to the formation of acid rain and smog (see Fig. 2.2).

The NO_x arises predominantly from the combustion of nitrogen compounds in the coal such as pyridine (so-called 'fuel-NO_x') but some are also formed by the chemical combination of the nitrogen and oxygen of the flue gases at the high combustion temperatures used (so-called 'prompt NO_x'). While the technologies exist for the removal of SO_2 and NO_x from the flue gases of power stations operating with coal combustion (SO_2 scrubbing and selective catalytic reduction (SCR) respectively), these are not always

Fig. 2.2 Nelson's Column, London, during the great smog of December 1952. *(Photo by N.T. Stobbs.) (https://commons.wikimedia.org/wiki/File:Nelson%27s_Column_during_the_Great_Smog_of_1952.jpg)*

practiced as they add significantly to the selling price of the electricity produced. Further, these methods cannot be applied when coal combustion is used for domestic heating purposes.

The hydrogen, nitrogen and sulphur contents of coal are due to the presence of a wide variety of chemical compounds arising from the original biomass, many of them aromatic in character and including a complex mixture of heterocyclic compounds related to pyridine and thiophene; coal also has a significant oxygen content, this being largely associated with compounds such as pyrroles and phenols. Because of these components, coal was used, prior to the increased availability of crude oil, as the source of many important chemicals (see Fig. 2.3). Coal was also used until relatively recently for the production (by pyrolysis) of Towns' Gas used for heating and lighting purposes in many parts of the world. It was also used as the source of the hydrogen for ammonia production and of the syngas used in the

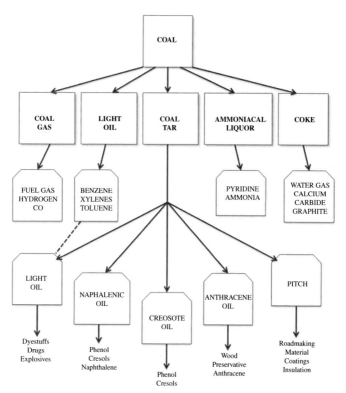

Fig. 2.3 Some chemicals produced by the destructive distillation of coal at about 1100°C.

Fischer-Tropsch Process operated in South Africa by Sasol during Apartheid; this entailed the operation of Lurgi gasifiers of the type developed in Germany in the period leading up to the Second World War. Ammonia synthesis and Fischer Tropsch Synthesis will be discussed further in a later section.

Coal for power generation and the steam engine

Coal was for many centuries used for heating purposes. However, its use in providing energy resulted from the development of the steam engine, the introduction of which as a source of power for machinery being responsible for the industrial revolution discussed in Preface. The earliest use of these steam engines was in the UK's cotton and woollen industries of Lancashire and Yorkshire; this development led to the first significant increases in greenhouse gas emissions discussed in Preface. Steam engines were soon thereafter used in railway engines (Fig. 2.4).

Coal for electricity generation

With the introduction of electrical power in the early 1800s, the steam engine was also used for the generation of electricity. The later development of the more efficient steam turbine led to the type of power station to which

Fig. 2.4 A typical steam engine (North Yorkshire Moors Railway, UK).

Fig. 2.5 A modern coal-burning power station: Moneypoint Power Station, Co. Clare, Ireland. *(With kind permission of ESB Archives.)*

we are all now accustomed. Fig. 2.5 is an aerial view of a typical modern power station used largely for coal combustion, showing the large area used for coal storage.

Fig. 2.6 is a schematic depiction of a typical cogeneration system that uses a primary gas turbine system coupled with a secondary steam turbine, this arrangement giving much higher all-over efficiency than one or other of these technologies alone. As we will see in subsequent sections, the use of coal as a fuel in such power stations has now almost completely been replaced

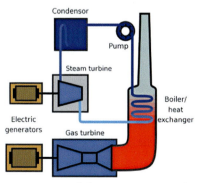

Fig. 2.6 Schematic representation of the arrangement of a typical cogeneration power plant. *(Source: Wikipedia.)*

by the combustion of natural gas or oil but the steam turbine is still the basis for electricity generation. Natural gas is now the fuel of choice as the emission of CO_2 is much lower than with the use of either coal or oil, as long as there are no serious emissions of methane during its transportation to the power station.

A turbine operating on its own would liberate a significant amount of energy as hot flue gases. With cogeneration, the hot flue gases are used to generate steam that is fed to a steam turbine that also produces electricity, adding to the all-over efficiency of the system. Some of the heat energy produced can be used for local heating systems.

It was not until the mid-1900s that it was generally recognised that coal combustion is associated with serious emissions of SO_2 and NO_x (mixtures of the oxides of nitrogen, predominantly NO and NO_2) and that these emissions cause serious health problems as well as bringing about the formation of 'acid rain', responsible for very significant damage to forestry. In consequence, clean air legislation has gradually been introduced to reduce these emissions and also those from domestic heating sources. With such legislation, the general use of smokeless fuel has become mandatory.

However, the greatest problem now associated with coal-burning power stations is the emission of greenhouse gas CO_2. For this reason, many coal-fired power stations have been converted to the use of either oil or natural gas since the amount of CO_2 emitted for a given amount of energy is thereby significantly reduced (see Box 2.1).

The proportion of coal used in the provision of energy varies significantly from country to country, depending very markedly on the availability of alternative sources of energy. As an example, Fig. 2.7 shows the use of coal and of various other sources of energy for electricity production in the US for the period 1950 to 2019. It can be seen clearly that the use of coal in the US since about 2000 has decreased steadily relative to that of other energy sources. The US Energy Information Administration expects that the use of coal in the US for electricity generation will decrease from 24% currently to about 13% in 2050. Similar decreasing usage patterns are expected for many other countries. However, China, India and Australia are still using increasing quantities of coal since these countries do not have a significant supply of alternative fuels. China, India and Australia are therefore currently very major contributors to greenhouse gas emissions.

It is of interest to examine the current use of electricity in some of the world's major economies. Considering first the United States, Fig. 2.8 shows the breakdown of electricity use for 2019. It can be seen that domestic use

BOX 2.1 Relative enthalpies of combustion of coal, n-octane and natural gas

The use of coal, oil and natural gas combustion for power stations give rise to different levels of the emission of CO_2. If we approximate the case of coal using the data for pure carbon and oil by a hydrocarbon with the data for n-octane ($n = 10$), we obtain the following data:

$$C + O_2 \rightarrow CO_2; \Delta H^\circ = -393.7\,kJ\,(C\,atom)^{-1}$$

$$C_8H_{18} + 12.5O_2 \rightarrow 8CO_2 + 9H_2O; \Delta H^\circ = -5104\,kJ\,mol^{-1}$$

$$\text{or } \Delta H^\circ = -638\,kJ\,(C\,atom)^{-1}$$

$$CH_4 + 2O_2 \rightarrow CO_2 + 2H_2O; \Delta H^\circ = -800\,kJ\,(C\,atom)^{-1}$$

It can be seen that the quantity of heat liberated per C atom combusted increases in the order: $C < C_8H_{18} < CH_4$ and that methane is the most effective fuel in terms of CO_2 emission. When the actual fuels used are considered rather than C and n-octane and the figures are now given as the emissions in kg CO_2 per kWh or per GJ energy produced (see Table 2.3), it again can be seen that natural gas combustion is significantly preferable to the combustion of coal or of oil as long as methane losses on the way to the plant are low.[b]

The data in Table 2.3 also show that the combustion of wood is only justifiable if sustainable sources are used. It can also be seen that peat and lignite are particularly undesirable energy sources and this is reflected in the fact that Ireland has closed its peat-burning stations and that Germany has completely ceased the use of lignite as a fuel.

Table 2.3 Emissions of carbon dioxide from various fuels as a function of the power generated.

Fuel	Emissions in kg CO_2/kWh	Emissions in kg CO_2/GJ
Wood	0.39	109.6
Peat	0.38	106.0
Lignite (Rhineland)	0.41	114.0
Hard Coal	0.34	94.6
Fuel Oil	0.28	77.4
Gasoline	0.25	69.3
Diesel	0.27	74.1
Liquid Petroleum Gas	0.23	63.1
Natural Gas	0.20	56.1

[b]Data from www.volker-quaschning.de.

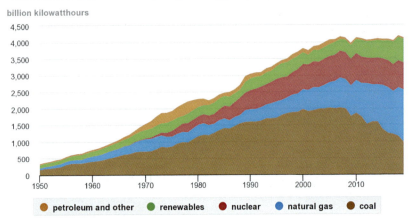

Fig. 2.7 Fuels used for electricity generation in the US from 1950 to 2019. *(www.epa.gov.)*

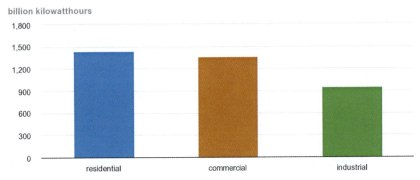

Fig. 2.8 US electricity usage. *(www.epa.gov.)*

was the most significant sector (about 40%) and that this was closely followed by the use by commercial enterprises, the third most significant user being the industrial sector.

The use of electricity for domestic purposes in the US is further broken down in Fig. 2.9. Approximately 50% of the usage can be attributed to

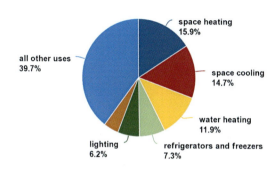

Fig. 2.9 US domestic electricity usage. *(www.epa.gov.)*

heating, cooling, water heating and refrigeration. While such usage will vary significantly from region to region, depending particularly on its geographical location, it can be seen that a very significant proportion of the US national energy usage is therefore devoted to maintaining comfortable living conditions.

Equivalent data for the European Union for 2017 show that 24.7% of the energy used in the residential sector comes from electricity and that less than 30% of this is used for space and water heating and cooling purposes (Fig. 2.10); however, when the European data on the use of energy includes the use of other resources such as natural gas, the categories space and water heating account for 75% of the usage. It is clear that a significant difference between Europe and the US is that the use of electricity for cooling purposes in Europe is much less and that natural gas is much more generally used for domestic heating.

Ireland is rather an exception in the European scene as domestic heating there is fuelled largely by oil. This is related to the fact that there is a large rural population for whom natural gas is not available. Table 2.4 shows the energy sources used for Irish electricity generation for the period 2005 to 2018. The use of coal has remained very steady, reflecting the use of the large Moneypoint plant shown in Fig. 2.5. (The much lower figure for 2018 reflects a problem that year with the turbines of the plant.) A large proportion of the coal burnt in Moneypoint has come from an open-cast mine in Columbia.

32 Sustainable energy

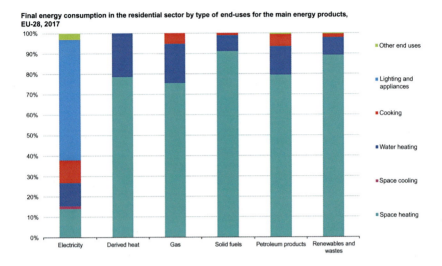

Fig. 2.10 EU domestic energy usage. *(https://ec.europa.eu/eurostat.)*

Table 2.4 Irish energy sources used for electricity generation.

Year	Coal	Peat	Oil	Wastes	Gas	Renewables	Imports
2005	549	211	287	0	995	161	176
2006	506	184	244	0	1186	213	153
2007	473	187	165	0	1330	240	114
2008	442	237	147	0	1438	309	39
2009	344	226	79	0	1402	353	66
2010	306	187	52	0	1558	321	40
2011	339	183	20	0	1327	466	42
2012	432	210	20	4	1216	452	36
2013	368	196	16	6	1129	484	193
2014	340	215	22	6	1087	550	185
2015	419	217	35	6	1064	676	58
2016	404	199	25	6	1318	646	0
2017	313	186	12	14	1348	764	0
2018	185	180	12	26	1377	877	0

Energy generated per fuel type; figures are in kton equivalents (ktoe).
Source: https://www.seai.ie/publications/Energy-in-Ireland-2018.pdf.

Natural gas is the major source of energy for Irish electricity generation and oil usage has almost completely stopped.[c] The use of renewable methods for electricity generation has also increased steadily; this is a topic to which we will return in Chapter 3.

There have also been a variety of other important uses of coal and a number of these still continue. The use of pyrolysis products for the production of important chemicals now produced from oil has already been mentioned; see Fig. 2.3. Town Gas production was also a very significant use and the gas produced was for many decades used for heating purposes. Prior to the widespread use of electricity, gas was also used for lighting. The use of Town Gas has now largely been replaced by natural gas. Some aspects of the use of natural gas will be discussed in a later section.

Coal use in cement production

Historically, another very significant consumer of coal has been the cement industry. Cement production (see Box 2.2) gives rise to the emission of large

---

BOX 2.2 Cement production

Cement has a long history. The Egyptians used cement made by calcining gypsum while the Greeks and Romans made their cement by calcining limestone to give lime and then adding either sand, to make mortar, or a mixture of sand and gravel, to give concrete. While the modern cement industry also depends on the decomposition of limestone (calcium carbonate), the chemistry is somewhat more complex. For example, the formation of an early form of Portland Cement was patented in 1824 by Joseph Aspdin, who produced it by firing a finely-ground clay together with limestone until the limestone had decomposed; the name Portland Cement was given since the resultant concrete looked like Portland Stone. Isaac Johnson made the first version of the modern Portland Cement by calcining a mixture of chalk and clay at high temperatures, 1400–1500°C, this giving a 'clinker' that has to be pulverised before the addition of other additives and subsequent use. Further developments followed such as the use of rotary kilns instead of the earlier vertical shaft kilns and the addition of gypsum to control the setting properties of the cement. The interested reader should further examine the fascinating history of the development of modern cements by carrying out a suitable Google search.

---

[c] Moneypoint now uses natural gas for most of the time.

34 Sustainable energy

quantities of CO_2, both from the burning of the fuel used in this high-temperature process and from the decomposition of the calcium carbonate component of the cement formulation involved in the all-over chemical process. It has been estimated that approximately 8% of the world's carbon dioxide emissions result from cement manufacture. A brief outline of the history of cement production is given in Box 2.2; the development of Portland cement was closely associated with the emergence of the Industrial Revolution since a high-strength cement resistant to aging in water was required to build the light-houses essential to a growing shipping trade.

If we consider only the decomposition of calcium carbonate:

$$CaCO_{3(s)} \rightarrow CaO_{(s)} + CO_{2(g)}$$

the standard enthalpy change, $\Delta H°$, is $178.3\,kJ\,mol^{-1}$. If coal is used as the fuel and we consider it to be entirely composed of carbon, the enthalpy of its combustion is $393.7\,kJ\,mol^{-1}$ (Box 2.1). Hence, if the combustion process is 100% efficient, an additional 0.45 molecules of CO_2 are formed by the heating process in addition to that formed in the decomposition of the $CaCO_3$. The efficiency of the heating process is far from 100% and so the ratio obtained in practice is well above 1.5. It is clear that using either oil or natural gas will improve the efficiency of the process. However, even if natural gas is used, cement production is one of the most serious contributors to the Greenhouse Effect. It has been estimated that anything between 8 and 10% of the world's annual emissions of CO_2 is emitted as a consequence of the production of various types of cement. The annual world production of cement exceeds 3000 million tons per year, the main producers being China (ca. 1800 million tons per year), India (ca. 220 million tons per year) and the US (ca. 63.5 million tons per year). A total of almost $900\,kg$ of CO_2 are emitted for every $1000\,kg$ of Portland Cement produced. For this reason, much research is currently being carried out on improved methods for producing cement: the use of different raw materials and additives; and modifications of heating methods used for the calcination process. We will return in Chapter 8 to a discussion of some developments in the methods used for cement production aimed at decreasing its carbon footprint.

Coal usage in iron and steel production

A further significant contributor to global CO_2 emissions is the iron and steel industry, this contributing between 4% and 7% of annual emissions. Table 2.5 shows the production figures for 2018 for the principal manufacturing countries and also includes, for comparison purposes, the data for the UK. The relatively low values for the UK are included because of its very significant contribution of developments in the iron and steel industries to the Industrial Revolution of the 18th Century; it is clear that while the UK was once at the forefront of iron and steel production, the major producers are now China, India and Japan, followed by the United States, South Korea and Russia.

A very clear description of the history of the production of iron and steel is to be found in an article entitled 'The Entire History of Steel' by Schifman in the magazine Popular Mechanics.[d] Schifman traces the discovery of metallic iron alloyed in meteorites and goes on to describe the emergence of the iron age, with cast and wrought iron and the importance of the incorporation of small amounts of carbon in the iron during the reduction process. Around 400 BC, Indian metal workers developed the use of a crucible to smelt a mixture of iron bars and charcoal pieces, heating this in a furnace using bellows to increase the temperature and producing the first steel; Indian steel was exported internationally. Early Japanese smiths also contributed to the development of modern steel by manufacturing intricate samurai swords. The use of coal addition to the iron ore during the smelting process was developed in the UK in the early 1700s and Sheffield became the centre of the steel production industry, using crucibles to smelt the iron ore-coal mixture. The blast furnace was invented by Henry Bessemer in the UK in 1856. The iron and steel industry in the US expanded

Table 2.5 Major crude steel production countries.

Country	Tonnage/million tonnes
China	928.3
India	106.5
Japan	104.3
United States	86.6
South Korea	72.5
Russia	71.7
Germany	42.4
Turkey	37.3
Brazil	34.9
UK	7.3

Steel production figures for 2018.
Data from www.worldsteel.org.

[d] https://www.popularmechanics.com/technology/infrastructure/a20722505/history-of-steel/.

Fig. 2.11 A Bessemer furnace in operation in Youngstown, Ohio, in 1941. *(Photo: Alfred T. Palmer Credit: Library of Congress, Prints & Photographs Division, Farm Security Administration/Office of War Information Color Photographs.)*

significantly after the Civil War using the Bessemer process (see Fig. 2.11). One of the pioneers in the new industry was Andrew Carnegie who helped establish the US iron and steel industries as the most important in the world. By 1900 the US was producing more than 11 million tonnes of steel a year, more than the UK and German industries combined. With the development of methods for producing stainless steel, US steel production was gradually overtaken by Japan and China.

As discussed further in Chapter 8, two main methods are currently used for the production of iron and steel: production from the ores (the so-called 'Integrated Route') and the reuse of scrap iron and steel ('Recycling Route'). The integrated route involves: (a) production of coke from coal by pyrolysis at 1000–1200°C in the absence of oxygen to drive off volatile components; (b) production of pig iron in a blast furnace at about 1200°C in which air is fed to a mix of iron ore and coke (plus 'fluxes' such as limestone to collect impurities), the coke forming the CO needed to reduce the ore; and (c) mixing of the pig iron with coke and up to 30% scrap, together with a small amount of flux, and heating using an oxygen lance to about 1700°C to produce molten steel. About 70% of the global production of steel involves

Fig. 2.12 Blast furnaces at Koninklijke Hoogovens plant at Ijmuiden, NL. *(Photo by J. Schoen, http://members.lycos.nl/fotoarchiefvon/hoogovens.JPG.)*

this integrated route. An example of a blast furnace facility is shown in Fig. 2.12. Such a plant gives rise to very significant emissions of CO_2 (Box 2.3).

The production of steel by recycling routes utilising scrap metal generally uses an electric arc furnace. As metallic iron is the raw material (rusted metal

BOX 2.3 CO_2 emission from steel production

Around 770 kg of coal is used to produce one tonne of steel from iron ore. Iron ore is commonly found as magnetite (Fe_3O_4), hematite (Fe_2O_3), goethite (FeO(OH)), limonite (Fe(OH)·nH_2O) or siderite ($FeCO_3$). For magnetite, the reduction by coke or coal can be represented by the following all-over reaction:

$$Fe_3O_4 + 2C \rightarrow 3Fe + 2CO_2; \Delta H° = +330 \, kJmol^{-1}$$

As indicated in the main text, in this reaction the carbon is first converted to CO and this CO is used as reductant but the all-over chemical reaction is as shown. It is clear that not only is a significant quantity of CO_2 formed in the reduction process but that CO_2 is also formed by the combustion of additional carbon in order to provide the all-over enthalpy change required for the reduction and also to provide energy to heat the system to the reaction temperature.

38 Sustainable energy

can also be used), much less energy is required (about 80% less) and the process also therefore produces much less CO_2. The production of steel by recycling is limited by the availability and purity of the scrap metal supply. About 20% of all steel was produced by recycling in the 1970s and that amount has increased to about 40% today, this figure including the scrap put into the feed for the integrated route described above.

Crude oil

Just as previous generations relied to a very great extent on the use of coal, much of our current way of life is dependent on the use of crude oil and its derivatives. This section starts by discussing the global reserves of crude oil, showing that it is a finite resource, and it then goes on to consider its current importance.

Table 2.6 shows the estimated oil reserves (in hundreds of million tons), the predicted lifetime of these reserves based on the production at the end of 2018, and the percentage of the total world reserves of each region of the world (figures in bold); it also shows the equivalent data for each significant oil-producing country in each region. The Middle East region has the largest estimated reserves, just short of 50% of the worldwide reserves, and also the largest rate of production. These figures are divided over a number of very significant oil-producing countries in that region. It is interesting to note that Canada is shown as having almost 10% of the world's reserves but it should be recognised that the majority of these are in the form of shale oil which is not currently exploited to any significant extent. A similar situation is found for Venezuela for which a significant proportion of its shale oil reserves, located in the Orinoco Belt, are also as yet unexploited; as a result, the predicted lifetime of the Venezuelan reserves is shown as being greater than 500 years. Of the Asian Pacific region, only China has any significant reserves but these are much less than the country's needs. What is particularly significant for the oil-producing countries of the world (with the exception of Venezuela, as noted above) is that the predicted lifetime of the known reserves in almost all cases is significantly less than 100 years. Even before the current concern regarding global warming, there had been a realisation that our current dependence on oil as a fuel cannot continue. Indeed, the Association for the Study of Peak Oil has estimated that the peak in the recovery of all liquid hydrocarbons is already past and that the reserves will be very much lower by 2050. As will be discussed further below, crude

Traditional methods of producing, transmitting and using energy

Table 2.6 Global crude oil reserves.

Region/ Country	Reserves/100 million tons	Reserves/ Production (R/P)	% share of World Reserves
North America	**35.4**	**28.7**	**13.7**
Canada	27.1	88.3	9.7
Mexico	1.1	10.2	0.4
United States	7.3	28.7	3.5
South America	**51.1**	**136.2**	**18.8**
Venezuela	48.0	>500	17.5
Europe	**1.9**	**11.1**	**0.8**
Norway	1.1	12.8	0.5
CIS	**19.6**	**27.4**	**8.4**
Kazakstan	3.9	42.7	1.7
Russian Federation	14.6	25.4	6.1
Middle East	**113.2**	**72.1**	**48.3**
Iran	21.4	90.4	9.0
Iraq	19.9	87.4	8.5
Kuwait	14.0	91.2	5.9
Saudi Arabia	40.9	66.4	17.2
United Arab Emirates	13.0	68.0	5.7
Africa	**16.6**	**41.9**	**7.2**
Algeria	1.5	22.1	0.7
Libya	6.3	131.3	2.8
Nigeria	5.1	50.0	2.2
Asia-Pacific	**6.3**	**17.1**	**2.8**
China	3.5	18.7	1.5
Total World	*244.1*	*50.0*	*100*

Oil Reserves (R) and Reserves/Production (R/P) at the end of 2018 for the major oil-producing countries. R/P is equivalent to the predicted lifetime of these reserves in years based on the 2018 usage. Data extracted from BP Statistical Review of World Energy, 2019.

oil is the source of many of the important chemical products that we now rely on and it therefore appears totally irresponsible to squander our reserves by simply burning them for energy production.

Crude oil and its various derivatives contribute about 8% of the global greenhouse emissions associated with fossil fuels. The process of extraction and the transportation of the crude oil to the refineries, generally located remotely from the oil sources, contributes very significantly to this figure. The latter emissions have many different origins, from the venting or flaring

40   Sustainable energy

Table 2.7 Average carbon intensity for upstream crude oil production.

Country	Average carbon intensity—Well to refinery/gCO₂eq/MJ
Algeria	2.05
Venezuela	2.0
Canada	1.76
Iran	1.68
Iraq	1.68
USA	1.09
Russia	0.96
UAE	0.7
Qatar	0.64
Norway	0.52
Bahrain	0.48
Saudi Arabia	0.45

The average volume-weighted global carbon intensity is approximately $10.3\,\mathrm{g}$ CO_2eq/MJ.

of associated gases to the energy inputs associated with heavy oil extraction and upgrading and they include emissions generated during transportation. In an important article published in Science in 2018, a large team of American scientists having as its lead author Adam R. Brandt, based at Stanford University, have given results of their estimates of 'well-to-refinery carbon intensity' values for all the major oil fields currently active globally.[e] They found that the well-to-refinery greenhouse gas emissions for 2015 were approximately 1.7 Gt CO_2 equivalent, this figure corresponding to about 5% of the total global fuel combustion emissions for that year. Table 2.7 summarises their data, showing the average carbon intensity for oil production by some of the more significant countries included in the survey. The global average value of the carbon intensity is ca. $10.3\,\mathrm{g}$ CO_2 eq/MJ.

It can be seen that the carbon intensity values given vary quite considerably from one oil-producing country to another. It is important to recognise that the emissions considered are not all of CO_2 but include emissions of methane (34% of the total CO_2 equivalent emissions); as was discussed in Chapter 1 the greenhouse effect of methane is approximately 25 times that of CO_2. Oil wells that vent natural gas, therefore, contribute more CO_2 equivalent than do wells

---

[e] https://www.osti.gov/pages/servlets/purl/1485127.

that flare the natural gas. Flaring contributes about 22% of the global volume-weighted upstream carbon intensity listed in Table 2.7, the counties contributing most to flaring include Iran and the US. It can be seen from the data of Table 2.7 however that the well-established oil producers of the Middle East are well below the global average. The authors of the Stanford study point out that a producer such as Saudi Arabia has a small number of extremely large and productive oil reservoirs and has low flaring rates; these wells also have low rates of water production and this helps minimise their emissions as the separation of any water requires significant amounts of energy. At the other end of the spectrum are Canada and Venezuela, both of which are producers of unconventional heavy oils the extraction of which is energy intensive and gives rise to significant CO_2 emissions.

Fig. 2.13 shows some of the steps involved in processing crude oil to usable products. The crude oil is first fed to a distillation column which separates the components into products ranging from gases to heavy residues. Each component is then further treated to give products ranging from fuel gas, used as a refinery fuel, to various components used for diesel and petroleum formulation and also to petroleum coke (also burnt to provide heat) and asphalt. Many of the units involved require significant amounts of energy: total emissions of CO_2 range from 1 million tonnes per year for a typical refinery to 3.5 million tonnes per year for a complex facility. (These high values fall only just below those for standard power plants.) According to the Greenhouse Gas Reporting Program (GHGRP) of the US EPA, the largest proportion of the CO_2 emission in US refineries comes from the various energy-producing combustion processes used (68%) while the majority of the remainder comes from the catalytic cracking and reforming units (27%); the remaining 5% comes from flaring and other sources. Very similar figures apply to the refineries in other parts of the world.

The majority of the products from an oil refinery are used for either transportation or heating purposes but a proportion is used as feedstocks for the petrochemicals industry. Many of the uses of the products of the oil refinery require the removal of all traces of sulphur and other potentially contaminated products such as aromatic components. Hence, a large proportion of the units in a typical refinery are devoted to purification steps such as hydrotreating. Other units are devoted to processes such as reforming, fluid catalytic cracking and alkylation. Further information on these processes can be obtained from the references given in the caption to Fig. 2.13.[f]

---

[f] It should be noted that the greenhouse gas emissions from the various products from an oil refinery are generally given separately from figures for transportation or energy production; however, the total well-to-wheel figures that are given in the literature relating to the transport sector include them.

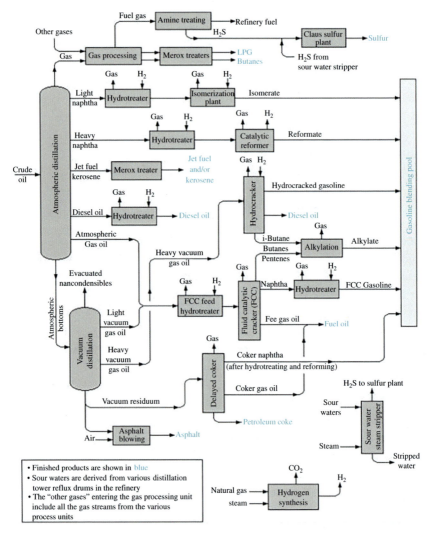

Fig. 2.13 Processes in a typical petroleum refinery. *(http://en.wikipedia.org/wiki/File: RefineryFlow.png. See also Contemporary Catalysis, Julian R. H. Ross, Elsevier 2018, ISBN: 9780444634740.)*

Natural gas

Natural gas is another important contributor to our current existence and it too has finite reserves. This section starts by outlining the global reserves of natural gas and then goes on to discuss some of the processes that depend on its widespread availability.

Traditional methods of producing, transmitting and using energy **43**

Table 2.8 Natural gas reserves.

Region/Country	Reserves as at end of 2018 /trillion m^3	Reserves/Production (Expected lifetime/yrs)	% Global share
North America	**13.9**	**13.2**	**7.1**
Canada	1.9	10.0	0.9
U.S.	10.0	14.3	6.0
South and Central America	**8.2**	**46.3**	**4.2**
Venezuela	6.3	190.7	3.2
Europe	**3.9**	**15.5**	**2.0**
Netherlands[a]	0.6	18.2	0.3
Norway	1.6	13.3	0.8
Ukriane	1.1	54.9	0.6
U.K.[a]	0.2	4.6	0.1
CIS	**62.8**	**75.6**	**31.9**
Azerbaijan	2.1	113.6	1.1
Russian Federation	38.9	58.2	19.8
Turkmenistan	19.5	316.8	9.9
Uzbekistan	1.2	21.4	0.6
Middle East	**75.5**	**109.9**	**38.4**
Iran	31.9	133.3	16.2
Iraq	3.6	273.8	1.8
Qatar	24.7	140.7	12.5
Saudi Arabia	5.9	52.6	3.0
United Arab Emirates	5.9	91.8	3.0
Africa	**14.4**	**61.0**	**7.3**
Algeria	4.3	47.0	2.2
Nigeria	5.3	108.6	2.7
Asia Pacific	**18.1**	**50.9**	**9.2**
Australia	2.4	18.4	1.2
China	6.1	37.6	3.1
Indonesia	2.8	37.7	1.4
Malaysia	2.4	33.0	1.2

[a]The Netherlands and the UK have relatively small reserves but are included because they are currently significant producers of natural gas.
Data for the countries with the predominant reserves in each region. (Figures for the end of 2018.)

The data given in Table 2.8 give the estimated natural gas reserves for some of the major gas-producing regions of the world. The global reserves are distributed throughout all the geographical regions but the majority of the individual national reserves are concentrated in countries of the Middle

Table 2.9 Natural gas composition—before and after processing.

Component	Before processing/%	Pipeline—after processing/%
CH_4	78.3	92.8
C^{2+}	17.8	5.44
N_2	1.77	0.55
CO_2	1.15	0.47
H_2S	0.5	0.01
H_2O	0.12	0.01

Data from Bradbury et al. (Footnote 11).

East and the CIS. The values of these reserves are traditionally quoted in cubic metres. The total yearly world production rate at the end of 2018 was 196.9 trillion cubic metres (final row of Table 2.8), this being the equivalent of the energy content of 3326 million tonnes of oil. As indicated earlier in this chapter (see Box 2.1), if the combustion of natural gas is used as a source of energy, the emission of CO_2 is significantly lower than that resulting from the combustion of coal or even of oil. It should be recognised that, in common with the burning of coal or oil, the combustion of natural gas can give rise to significant proportions of NO_x due to the combination of nitrogen and oxygen of the air at the high temperatures of the combustion process (so-called 'prompt NO_x', as discussed under coal combustion). Natural gas as recovered is a mixture of methane and lower volatile hydrocarbons as well as N_2, CO_2, H_2S and H_2O; the quantities of these components vary considerably with source. Before the gas is piped to the consumer, a significant proportion of the impurities are removed. Table 2.9 gives data from a US. Department of Energy report that shows the average compositions of natural gas as recovered and after purification, these being compiled from data for 2012. As natural gas is odourless, small quantities of mercaptan (methanethiol) are added to the gas (at levels up to 10 ppm) before it is distributed for use in heating applications since the harmless but pungent odour of this component is easily detected at concentrations as low as 1.6 ppb; the combustion of the added mercaptan gives rise to the emission of only small quantities of SO_2.

As shown in Table 2.8, the United States currently possesses about 6% of the world's reserves of natural gas, this corresponding to about 13.9 trillion m^3 (500 trillion ft^3). This value is a relatively recent development, the known

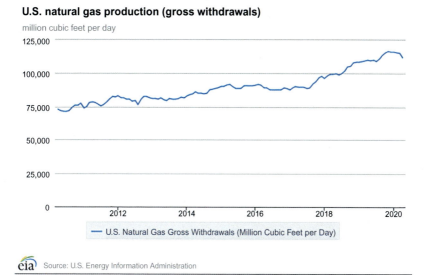

Fig. 2.14 The increase in US natural gas production since 2010. *(https://www.eia.gov/naturalgas/crudeoilreserves/.)*

reserves having more than doubled from a value of about 5.8 trillion m^3 (210 trillion ft^3) in 1978; the data of Fig. 2.14 show that this growth in production is still increasing continuously. This very large increase in production is due to the fact that much of the natural gas now produced in the United States is recovered by hydraulic fracturing, more commonly known as 'fracking' (and sometimes called 'fraccing'). Fracking has been practiced for some decades as a means of improving flows during the extraction of oil and gas from reservoir rocks. It entails the injection under pressure of a mixture of water, proppants (commonly sand) and chemical additives. The injection process fractures the rocks and the injected proppant maintains the size of the fissures, this enabling outwards flow of the oil or gas. The method was first used in 1947 to improve the flow of crude oil from oil wells. Most natural gas is now produced in the US using so-called 'horizontal slickwater fracturing' from deposits of shale using more water and higher pressures than used previously. The first of the wells using this method was put into production in 1998 at the Barnett Shale Field in North Texas. Since 1998, the number of such wells has increased very rapidly so that natural gas is now produced in most regions of the US. The fracking technique is also now applied in other countries such as Canada, Germany and the Netherlands and in the UK's gas fields in the North Sea.

46   Sustainable energy

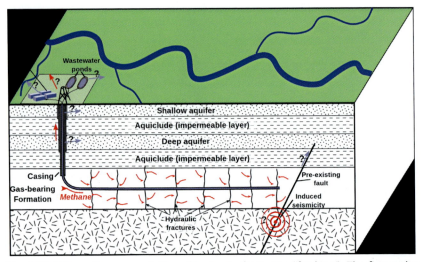

Fig. 2.15 Schematic representation of hydraulic fracturing ('fracking'). The figure also shows the routes for possible leakage of natural gas and fracking liquids into the environment. https://en.wikipedia.org/wiki/Hydraulic_fracturing.

As shown schematically in Fig. 2.15, the modern application of fracking involves the drilling of horizontal bore-holes in the gas-bearing shale bed. This shale layer may be at a depth of several thousand metres. The hydraulic fracturing process, brought about by the injection of the fracking fluid at high pressure, gives rise to a series of fissures that are connected to the bore-hole through which the gas passes to the surface. Although significant precautions are taken to prevent leakage of gases from the borehole, there is evidence of significant losses of methane to the atmosphere. There can also be appreciable leakage of the fracking fluid into the surrounding rock structures and also into aquifers as shown by arrows with a question mark of Fig. 2.15. There is in consequence a significant resistance to the introduction of the fracking technique in many countries, not only as a result of environmental concerns caused by this potential for leakage of natural gas but also by fears that fracking can cause earthquakes leading to the creation of additional leakage paths and structural damage at the earth's surface. Fracking is currently banned in several US states and also in a number of European countries including Ireland.

It is interesting to note that while the production and transportation of natural gas by more traditional means has a relatively small carbon footprint, this is not the case for natural gas produced by fracking as a proportion of the gas is used at each stage of the recovery, processing, transmission, storage and distribution stages. A relatively recent report [11] shows that the proportion of

Table 2.10 Proportions of US natural gas used prior to delivery.

Stage of production	Natural gas volume/ Billion cubic feet	% of Total production
Total Production	**26,826**	**100**
Production Stages	1283	4.7
Processing	1224	4.6
Transmission and Storage	842	3.1
Distribution	66	0.24
Total Losses	*3415*	*12.7*
Quantity to Consumers	**23,411**	**87**

the natural gas produced in the US that reaches the final consumer is only 87%; see Table 2.10. The total CO_2 emissions for 2012 associated with the 13% of the natural gas used in production and distribution were 163.7 million metric tonnes while the associated emissions of unburnt methane were 154.7 million metric tonnes CO_2 equivalent; the total CO_2 emissions associated with the end use were 1234 million metric tonnes.

Concluding remarks

This chapter has given a brief introduction to the widespread use of fossil fuels - coal, oil and natural gas - for energy-related purposes, all of these having been introduced before the emergence of the current state of awareness of the problems of greenhouse gas emissions. Each of these traditional uses gives rise to large quantities of emitted CO_2. The advantage of using natural gas as opposed to oil or coal as an energy source is very significant due to the lower emissions of CO_2 for a given amount of energy produced. Nevertheless, unless the CO_2 produced can be collected and used or safely stored (Chapter 4), the emissions from natural gas combustion are still very substantial. Chapter 3 now examines some of the ways in which the energy-related CO_2 emissions discussed in this chapter can be reduced by the introduction of new 'renewable' technologies. Chapter 4 then examines the production of 'grey' and 'blue' hydrogen from natural gas and of 'green' hydrogen by the electrolysis of water using 'renewable' electricity, a topic considered in more detail in Chapter 8. Chapter 4 also discusses the important role that the hydrogen produced plays in many different processes, including the synthesis of ammonia and of methanol.

CHAPTER 3

Less conventional energy sources

Introduction

Having considered in Chapter 2 the production of energy from unsustainable fossil fuels as well as some of the uses of these fossil fuels as feedstocks for the chemical industry, we will now consider the production of energy from more sustainable sources. In this context, the use of the word 'sustainable' is used to indicate that these sources are either totally renewable or that the method of supply makes significant use of renewable sources.

The sun emits light and other radiation at a rate of 3.846×10^{26} watts. This energy is formed as a result of nuclear reactions occurring within the sun's mantle and so the use of this energy has a negligible consequence on subsequent supplies. As shown schematically in Fig. 3.1, the intensity of the radiation reaching the earth's atmosphere is 1368 watts per square metre (W/m^2); of this, some is absorbed by the atmosphere and the average density at the earth's surface on a surface perpendicular to the rays is approximately $1000\,W/m^2$. This quantity is far in excess of that required for the world's energy needs.[a]

Fig. 3.2 shows the relative contributions of CO_2 equivalent emissions of each of the most important forms of 'renewable energy' relative to those of coal and gas (Chapter 2), these data having been calculated in 2014; the data take into account the emission of any greenhouse gases formed during the production of the energy, for example by the operation of pumps and other equipment.

Most of the routes that will be discussed below depend in some way or another on the radiation reaching the earth from the sun, the only exceptions being the use of nuclear energy and of those that make use of energy stored in the earth's crust (geothermal energy) or formed

---

[a] It has been shown that radiation density is such that if all the energy falling on the US state of Texas at noon could be converted to electricity, the amount produced would be approximately 300 times the total amount of electricity generated by all the earth's power plants.

Sustainable Energy
https://doi.org/10.1016/B978-0-12-823375-7.00001-9

Copyright © 2022 Elsevier B.V.
All rights reserved.

49

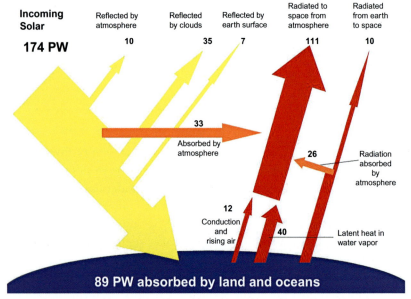

Fig. 3.1 Uptake and re-emission of solar power by the earth's surface. *(From https://en.wikipedia.org/wiki/Solar_energy)*

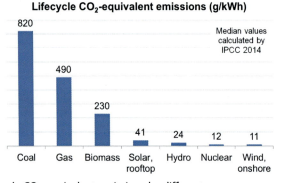

Fig. 3.2 Life-cycle CO_2-equivalent emissions by different energy sources. *(From https://www.ipcc.ch/report/ar5/syr/)*

as a consequence of gravitational forces between the earth and the moon (tidal energy).

Nuclear energy

There are three types of nuclear reactions, all of which create energy: fission, radioactive decay and fusion. Only fission is currently used for the generation of energy; this is described further below after a brief outline of each of the processes.

Nuclear Fission occurs in the nuclei of elements such as uranium and plutonium. Fission is a process in which the nucleus of the element breaks down completely to form lower atomic weight elements. This is a chain reaction initiated by bombardment with neutrons to bring about the fission of the nucleus with the concurrent emission of additional neutrons. The neutrons formed can then initiate a further reaction, thus resulting in a nuclear chain reaction. The reaction therefore has to be very carefully controlled, for example by inserting graphitic carbon moderator rods, in order to capture the excess neutrons and to prevent explosion (such as that in an atomic bomb). The reaction of ^{235}U (one of the naturally occurring isotopes of uranium, found at an abundance of 0.72%) gives a series of reactions, the first of which is as follows:

$$^{235}U + n \rightarrow {}^{140}Ba + {}^{96}Kr + 3n + \sim 200\,mev$$

The 200 mev liberated per fission event is equivalent to $19.2 \times 1012\,J$ per mol of the U compound decomposed, a figure far in excess of the energies associated with normal chemical reactions. The fission process is depicted schematically in Fig. 3.3.

Plutonium 239 (^{239}Pu) is also used as a nuclear fuel. This element occurs in nature but it can also be produced from ^{238}U by bombardment with neutrons. Yet another fissionable isotope, ^{238}Pu, can be produced by bombarding ^{238}U with deuterons (deuterium atoms).

Radioactive decay occurs naturally: all radioactive elements gradually break down to give elements of lower atomic weight; this decay creates lower levels of energy and so the processes based on decay are only used in very specialised situations. Both fission and decay generally give rise to radioactive products and the storage and disposal of the radioactive waste is a very significant barrier that has led to the banning of the introduction of nuclear reactors in a number of countries, such as Ireland, or their phase-out, as in Germany.

Nuclear fusion involves the reaction of the nuclei of two lighter atoms to give a heavier element, the process creating significant amounts of energy (see Fig. 3.4). Fusion is an energy-creating technology which has not yet been commercialised although much research continues to be performed on the topic. The reaction is generally carried out with one of the heavier isotopes of hydrogen (deuterium or tritium) and requires very high temperatures such as those found in plasmas, tens of millions of degrees.[b]

---

[b] A hydrogen bomb using fusion achieves the required temperatures by the incorporation of a fission component.

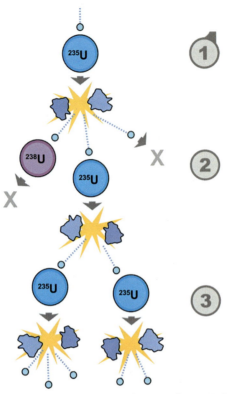

Fig. 3.3 Schematic representation of the fission of 235 U brought about by bombardment with neutrons, showing the chain reaction that occurs as a result of the creation of more and more neutrons. At each collision of a neutron with a uranium species, more than one additional neutron is emitted and so the reaction speeds up, ultimately giving an explosion, unless some of the neutrons are absorbed by other moderating species. *(From https://en.wikipedia.org/wiki/Nuclear_fusion)*

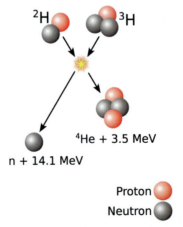

Fig. 3.4 The fusion reaction of deuterium (2H) or tritium (3H) to give helium (4He) plus large amounts of energy. *(From https://en.wikipedia.org/wiki/Nuclear_fusion)*

Fig. 3.5 The nuclear reactor at Calder Hall in Cumberland (UK) has been in operation since 1956. *(From https://en.wikipedia.org/wiki/Sellafield)*

The advantage of using such a process would be that there would be no radioactive waste. However, despite much research on the topic, no reactor has yet been developed to allow commercial operation since the energy input required to reach reaction temperature and pressure using currently available equipment is of the same order as the amount of energy produced. A further problem is that the neutrons emitted in the process degrade the reaction vessel with time and so the development of nuclear fusion will also require major innovations regarding construction materials.[c]

The Use of Nuclear Power. The first commercial reactor employing nuclear fission was that at Calder Hall in the United Kingdom (Fig. 3.5), first connected to the grid in 1956. Shortly after, several plants were commissioned in the United States and a significant number of plants were thereafter built in many other countries. France is a particularly noteworthy user of nuclear electricity since much of its demand is now satisfied by nuclear generation; as a result, France is also a very large exporter of electricity. The steady growth of France's nuclear electrical production is shown in

---

[c] The processes of fission, decay and fusion are well described on the web; see for example https://en.wikipedia.org/wiki/Nuclear_power.

54 Sustainable energy

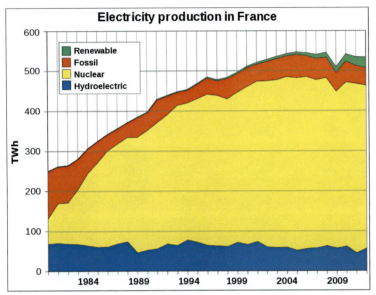

Fig. 3.6 The significant growth of the proportion of French nuclear electricity generation since 1980. *(From https://en.wikipedia.org/wiki/Nuclear_power_in_France)*

Fig. 3.6; French production is currently about 400 terawatt hours (TWh). The total global generation of electricity by nuclear reactors was 2563 TWh in 2018, this being about 10% of worldwide electricity generation. The USA is now the main producer of electricity using nuclear power, having a 30% share of global production. In December 2019, there were 4433 civilian nuclear reactors worldwide, these having a combined electrical capacity of 395 gigawatts (GW). An expansion of almost 50% in this power output is either in construction or planned, this growth occurring predominantly in Asia. It is clear that the production of this amount of electricity has saved the emission of significant amounts of CO_2 in the period since nuclear production was first commercialised.[d] However, as indicated above, there are many countries where there is considerable opposition to the use of nuclear generation as there are major concerns regarding the safety of nuclear reactors and about the safe disposal of nuclear waste. In the United States and in the United Kingdom, an expansion of nuclear power is planned as part of the future energy supply, together with energy from renewable resources.

---

[d] It has been claimed that the emission of CO_2 has been reduced by 64 billion tons since the 1970s by the introduction of nuclear power.

BOX 3.1 Cold fusion

Great scientific excitement was generated in 1989 when Pons and Fleischmann reported that they had produced excess heat during an experiment involving the electrolysis of heavy water, D_2O, using Pd electrodes. They reported that they had observed small quantities of neutrons as well as tritium, the expected products of a nuclear fusion reaction. Attempts by many scientists to replicate this work led to many papers on the topic, some supporting the original claims and others not. Finally, a recognition emerged that the reported results were faulty and it became clear that Pons and Fleishmann had not observed the products reported. 'Cold Fusion' therefore died a natural death.

Any energy-producing technology depends on there being a significant supply of the fuel used. Uranium is a reasonably common element, having an abundance similar to those of tin or germanium. It is currently economically extracted from deposits that contain relatively high concentrations. However, it is also present at low concentrations in many rocks and also in seawater; extraction from the seawater is possible but the cost would be significantly higher than that from currently used sources. Reserves of the most economically extracted uranium-containing materials are currently sufficient for about a century at the current rate of use. While accepting that the costs of extraction could increase if the easily recovered reserves are depleted, it could be argued that nuclear energy is a renewable resource as the total amount of uranium used is very small compared with the total level of reserves. However, the currently unsolved problems associated with the safe disposal of nuclear waste make this a very questionable approach.

Another potential source of renewable energy, from Cold Fusion, now totally discounted, is described in Box 3.1.

Geothermal energy

Geothermal energy is energy that is derived from within the earths crust. Most of this arises from radioactive decay occurring deep within the earth's core but part is residual energy remaining there from the original formation of the planet. The temperature difference at the boundary between the molten core and the earth's crust can reach values of over 4000°C. The movement of components of the mantle towards the surface (e.g. in volcanoes) brings some of this energy towards the surface. The water from hot springs

Fig. 3.7 Kafla geothermal power station, Iceland. *(From https://en.wikipedia.org/wiki/Geothermal_power)*

heated by such volcanic activity has been used for centuries in thermal baths and some of the sources of hot water have been used in thermal power stations. Fig. 3.7 shows one such station in Iceland, a country in which there are many such sources of geothermal energy situated near tectonic plate boundaries by which the energy can more easily reach the surface. Whether the heat supplied by a particular source is sufficient to produce electricity or is only usable for local heating purposes depends on the temperature of the water. Table 3.1 shows the geothermal electric capacities of the countries producing the largest amounts of electricity using geothermal sources.

It can be seen from these data that although Iceland produces only 5% of the world's supply of geothermally-generated electricity, this represents 30% of the country's own electricity use. The third country on the list, Indonesia, has the largest estimated reserves in the world (28,994 MW) and it is predicted that this country will soon overtake the USA in thermal electricity production. The main operation in the USA is at the Geysers in Northern California.

The efficiency of geothermal power plants such as that shown schematically in Fig. 3.8 is low, ranging from 10% up to 23%. Whether or not the heat from such a source can be used to generate electricity depends greatly on the temperature of the source. In many locations where the source

Table 3.1 Installed geothermal electricity capacities of the major producing countries.

Country	Capacity in 2010/MW	% of national electricity production	% of global geothermal electricity production
United States	3086	0.3	29
Philippines	1904	27	18
Indonesia	1197	3.7	11
Mexico	958	3.0	9
Italy	843	1.5	8
New Zealand	628	10.0	6
Iceland	575	30.0	5
Japan	536	0.1	5

Source: Condensed from a larger listing in Wikipedia.

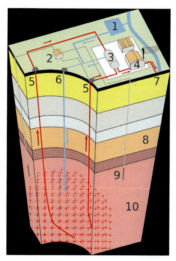

Fig. 3.8 Schematic representation of a geothermal electricity generation plant. 1, Reservoir; 2, Pump house; 3, heat exchanger; 4, Turbine hall; 5, Production well; 6, Injection well; 7, Hot water to district heating; 8, Porous sediments; 9, Observation well; 10, Crystalline bed rock. *(From https://en.wikipedia.org/wiki/Geothermal_power)*

temperature is too low, geothermal energy is used only for local heating purposes such as district heating and in greenhouses. In order to use hot aquifer sources (which can be at depths of up to 10 km) for electricity generation, the energy is extracted by a geothermal heat pump similar in operation to that shown in Fig. 3.9 of Box 3.2 so that temperatures high enough to operate the associated steam turbines are achieved. This is only economically

BOX 3.2 The operation of a heat exchanger for domestic heating purposes

Fig. 3.9 shows the operation of a domestic heat exchanger in which cold air is being delivered into a building for cooling purposes. The energy removed by the fan/evaporator system is taken up by the condenser which is then cooled either by the external air or by extraction to the surrounding ground. (This process is similar to that found in a domestic refrigerator but the latter operates without a fan.) By changing the direction of the operation of the fan, heat can be extracted from the exterior surroundings of the building (air or ground) and supplied via the condenser for heating purposes (as in so-called 'air-to-water' or 'ground-to-water' systems). Such central-heating systems are now in relatively common use; similar installations can also be used solely for ventilation purposes, with the transfer of the heat from the expelled air and its surroundings to the incoming air. Although such heat exchangers are generally most simply installed during the construction phase of a building, retrospective fitting is also now possible as a result of the development of suitable modules. Systems of this type require the building to be relatively 'leak free'; if not, more energy is required to circulate the air than is extracted from the surrounding air or ground and there is no reduction in greenhouse gas emission unless totally renewable electricity is used. Similar heat exchange units are also used in district heating systems recently introduced in a number of countries, particularly in Sweden. As with other forms of geothermal energy, the system is only 100% sustainable if renewable electricity is used for operating the system; if electricity derived from fossil fuels is used, the all-over generation of greenhouse gas emissions is generally reduced by only 60%–70%.

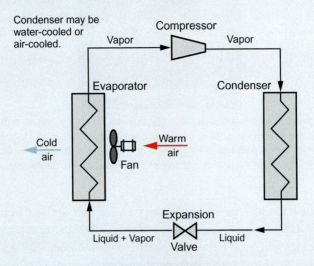

TYPICAL SINGLE-STAGE
VAPOR COMPRESSION REFRIGERATION

Fig. 3.9 The operation of a heat exchanger. *(From https://en.wikipedia.org/wiki/Heat_pump_and_refrigeration_cycle)*

feasible if the amount of electrical energy generated is significantly higher than that needed to operate the pumping and related systems. Not only does a plant such as that shown in Fig. 3.8 generate electricity but it also provides hot water for local district heating purposes.

Tidal energy

A number of different renewable technologies dependent on water rely on energy originally imparted by solar radiation, as discussed further below. However, one exception is tidal energy, this arising predominantly from gravitational forces generated by the moon as it orbits around the earth; gravitational forces from the sun also contribute to the magnitude of the tides but to a lesser extent. (The actual magnitude of each tide depends on the relative positions of the earth, the moon and the sun.) The presence of tides creates very large movements of water and these movements have associated with them very significant amounts of kinetic energy.[e] The dissipation of this energy does not occur evenly throughout the world's oceans and some areas have much higher tidal energies than others. Fig. 3.10 shows the global distribution of tidal energy; although a significant proportion of the total

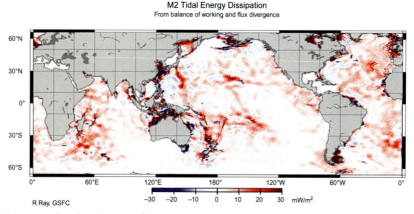

Fig. 3.10 The distribution of global tidal energy dissipation, showing those areas where tidal effects are greatest. *(From https://earthobservatory.nasa.gov/images/654/dissipation-of-tidal-energy.)*

---

[e] A tidal flow rate of 10 mph gives an energy output equal to or greater than that for a wind speed of 90 mph.

Fig. 3.11 A tidal mill on the west coast of Ireland. Water entered and left the lagoon visible in the background through the mill-race on the right-hand side of the main structure; the mill wheel was mounted in this mill-race which is connected to the sea through a man-made channel. *(Photo: J.R.H. Ross.)*

tidal energy available is dissipated in the middle of the oceans, there are coastal regions where the effects are very high and others where the effects are much lower.

The tidal movement has for many centuries been used as a source of mechanical energy. An example of a tidal mill on the west coast of Ireland that began operation in about 1804 is shown in Fig. 3.11. This mill was built at the entrance to a large tidal lagoon which has an area of about 0.5 km^2. Water was trapped twice a day at high tide (tidal differences up to about 4 m) and allowed to flow out through the mill-race (visible in the photograph to the right of the main structure) at each low tide. The mill was used very profitably for grinding wheat for sale in England. (At that time, England had been cut off from European markets by the Napoleonic wars.) The mill ceased production when power from other sources became more generally available at the beginning of the 20th century. This technology is very similar to that now used in hydroelectric power generation, a topic that will receive further attention below.

Less conventional energy sources 61

Fig. 3.12 The SeaGen tidal stream generator situated in Strangford Lough, Northern Ireland that operated from 2006 to 2019. *(From https://en.wikipedia.org/wiki/SeaGen)*

Tidal energy is currently used for the generation of electricity in a limited number of locations, many of them in regions with high tidal energies (Fig. 3.10). In most cases, this is achieved not by the creation of tidal dams but by the use of submerged turbines. The earliest example of the use of a large commercial turbine assembly is shown in Fig. 3.12. This system, recently decommissioned following successful operation for 11 years, was located at the narrow entrance of a large tidal area, Strangford Lough, and the power produced was used in the geographical region close to the generation system: in the neighbouring town of Portaferry and its immediate surrounds. (This turbine was switched off if there was a large movement of fish into or out of the Lough.[f]) Power generation in such a system is at a maximum for several hours on either side of high tide so operation lasts only about 10 h per day. Hence, either storage is required or the power supply has to be integrated into the regional electricity supply system, using power from another source in periods when generation does not occur. We will return to the important topic of storage of electrical energy in Chapter 7.

The amount of energy associated with the tidal flow can be very high, depending on the location. Tidal flows along the Atlantic seaboard of

---

[f] Environmental arguments generally preclude the construction of tidal barriers of the type formerly used for the mill shown in Fig. 3.11. The is a significant danger of silting up and the effects on wild life can also be very damaging.

62 Sustainable energy

Europe are among the highest shown in Fig. 3.10, with the tidal flows around the east coast of Ireland being particularly strong. A report from Sustainable Energy Ireland[g] indicates that the use of the practically accessible tidal resources around the Irish coast could provide over 5% of the electric requirements for the whole island. The theoretical limit is much higher (up to 5 times the total electricity consumption) but there is still a need for considerable developments in the technology required, much of which is in its infancy. Environmental constraints may also have an important role in determining whether or not the necessary developments would ever be carried out. Similar arguments apply to potential developments in the other regions with high tidal energies shown in Fig. 3.10.

A closely related area of interest is the use of the energy associated with marine currents, these being a consequence of the movement of the tides. The energies associated with marine currents can be very high and they can be utilised using turbine devices similar to those described above. It has been estimated that the total worldwide power in ocean currents is about 5000 GW, these corresponding to power densities of up to $15\,\mathrm{kW/m^2}$. An example of an ocean current that carries very high energies and that might be used for power generation is the Gulf Stream which flows past Florida and up the eastern coast of the United States. A serious obstacle to the development of such systems to harvest ocean currents is the distance of the currents to be utilised from land and the consequent need for long power cables. There are also potentially significant environmental problems. There do not appear to be any commercial examples of turbines using marine currents even though there has been a significant amount of research on the subject.

Wave power

Wave power, a closely related form of renewable energy, results entirely from solar radiation. Waves are created by the passage of winds, generated by solar radiation, passing over the surface of the ocean. The energy contained in the waves depends on many factors such as the wind speed, the depth of the water and the topography of the sea-bed. A moderate ocean swell in deep water some distance from a coastline with a wave height of 3 m and an average period of 8 s has energy approximately equivalent to 36 kW per metre of wave crest. In a major storm, with wave crests of about 15 m, the energy of a wave crest can rise to about 1.7 MW per metre. As shown in Fig. 3.13, some coastal regions of the world have higher wave

[g] https://www.seai/publications/Tidal_Current_Energy_in_Ireland_Report.pdf.

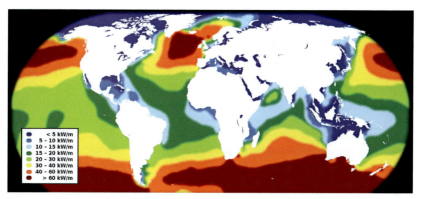

Fig. 3.13 Global distribution of wave energies. The figure shows wave energy flux in kW per meter wavefront. *(From https://en.wikipedia.org/wiki/Wave_power)*

energies than others; the most viable areas for wave power generation are on the western seaboard of Europe, the northern part of the UK, parts of the Pacific coastlines of both North and South America, Southern Africa, the southern coasts of Australia and New Zealand. Further information on wave energy and related topics can be found on the web, for example at https://en.wikipedia.org/wiki/Wave_power.

A series of different types of devices have been developed to utilise wave energy, for example using turbines of the type discussed above for tidal power or some type of floating device tethered to the ocean bed. As with tidal energy, these technologies are generally at a relatively early stage of development. An important aspect of wave energy is that it is not as consistent and predictable as is tidal energy. The use of wave energy also presents a range of environmental constraints similar to those associated with tidal energy.

Hydroelectric power

The three most important methods of generation of renewable energy from a practical point of view now in use are hydroelectric systems, wind turbines and solar generators and these will be considered in the following three sections. Hydropower has been used for several centuries. It was an important basis for the development of the industrial revolution when it was used to provide the mechanical energy required to operate textile mills (see Chapter 2). In the twentieth century, hydroelectric power generation became a very important source of renewable electricity in countries where there are large rivers in which such generating stations can be operated. One of the world's first large-scale generating stations was on the River Shannon in Ireland and this started generation in 1929 (see Box 3.3) but

BOX 3.3 The Shannon hydroelectric scheme

When Ireland became independent of the UK in the early 1920s, it was recognised that the country had very limited natural resources such as coal. Natural gas reserves were only found in 1973 (coming on stream in 1978) and so the new nation had been largely dependent on the use of large reserves of peat for heating purposes. (Peat was also used later in a limited way for electricity generation.) It was recognised that the River Shannon, the longest river in the British Isles, was a potential energy resource and this led to the creation of the Shannon hydroelectric scheme. The Ardnacrusha power station that was then built (Fig. 3.14) was opened in 1929. The plant was built by the German company Siemens-Schuckert and generated 85 MW of electricity, sufficient (after the construction of the appropriate grid system) for the requirements of the whole country. At the time, it was the largest hydroelectric power plant in the whole world. The new power plant also required the construction of a series of dams and bridges as well as the diversion of the majority of the water from the upper reaches of the Shannon and its associated lough system through a new canal leading to the plant, the water having an average head of 28.5 m. It is interesting to note that the cost (£5.2 million) was at that time an enormous sum, thus representing more than 20% of the budget of the newly created state.

Fig. 3.14 The Ardnacrusha power plant of the Shannon Hydroelectric scheme. *(Photograph reproduced with the kind permission of ESB Archives.)*

Continued

BOX 3.3 The Shannon hydroelectric scheme—cont'd

The diversion of the main flow of the Shannon that was carried out had enormous environmental consequences which at the time of construction were largely disregarded. One of the immediate consequences was that the reaches of the Shannon above Ardnacrusha, which had been one of the most important angling locations in the West of Ireland, were no longer accessible to breeding salmon. Navigation was also affected and there were problems of flooding upriver. Some of these difficulties have since been somewhat alleviated, for example by the construction of a fish pass for the migrating salmon, but flooding problems still persist and navigation is very difficult. The Shannon Hydroelectric scheme still supplies renewable electricity but its output now represents only about 2% of the country's requirements.

Table 3.2 Hydroelectric generation in the top producing countries in 2014.

Country	Annual hydroelectric production/TWh	% of total electricity production
China	1064	18.7
Canada	383	58.3
Brazil	373	63.2
United States	282	6.5
Russia	177	16.7
India	132	10.2
Norway	129	96.0

many larger-scale stations have since been built. In 2015, hydropower generated some 16.6% of the world's electricity requirements, this corresponding to more than 2/3 of all renewable electricity. Table 3.2 gives data for 2014 for the world's top hydroelectric generating countries, showing the % hydroelectric generation compared with total electricity production for each of these countries. China is top of the list and it would appear that its share of production is still increasing rapidly as a result of continued new construction. Canada, Brazil and, in particular, Norway all produce a significant proportion of their national demands of electric power from hydroelectric generation. Although the United States is a large producer, the proportion from this source is relatively low. Hydroelectricity is a very reliable source of energy as it is dependent in most cases on the creation of a reservoir of water above the power station and this ensures continuous generation under all but the most extreme drought conditions.

Fig. 3.15 The 124 m high Krasnoyarsk Dam (Siberia, Russia). *(From https://en.wikipedia.org/wiki/Krasnoyarsk_Dam)*

There have been significant environmental aspects of the creation of large dams for hydroelectric generation. As discussed in Box 3.3, damming the River Shannon had some significant effects, the most lasting of which has been continued flooding of the upper reaches of the river almost every winter. Another example of environmental problems is the microclimatic changes that occur in the Siberian city of Krasnoyarsk due to the construction, completed in 1972, of a large 124 m high concrete dam on the River Yenisey at Divnogorsk, 30 km upstream of Krasnoyarsk; see Fig. 3.15. This dam was created largely to provide energy for the large aluminium plant in Krasnoyarsk (at that time a closed city inaccessible to Westerners). When the electricity generation plant started operation, it supplied an enormous 6000 MW of power, a level only since exceeded by the Grand Coulee Dam in Colorado, USA, which reached 6181 MW in 1983. The completion of the Krasnoyarsk Dam resulted in the creation of a large lake, now known as the Krasnoyarsk Sea, this having an area of approximately 2000 km^2. (A fascinating aspect of this dam is that shipping using the River Yenisey can be transported past the dam using a very large electric rack railway specially constructed for the purpose.) The volume of water in the Krasnorarsk Sea is more than 70 km^3 and it passes the dam continuously during all seasons as a liquid despite the fact that the river would have previously frozen over during the winter months. This enables the year-round passage of shipping in the lower reaches of the river, something that was not possible before its construction. However, the outflow of cold water at all seasons of the year also affects the climate in the vicinity by causing cylindrical rotation of the weather in the river valley, the direction of rotation depending on whether the land is hotter or colder than the river. Krasnoyarsk therefore suffers from severe environmental problems since the emissions from its industrial activities are trapped in its vicinity by these unusual air vortices rather than being released into the surroundings.

Wind power

The generation of electricity using wind power is now familiar to everyone and land-based wind turbines such as that shown in Fig. 3.16 are common. Wind generation has increased very significantly over the last two decades (see Fig. 3.17) and additional capacity is added yearly, the main limitation to such expansion being the capacity of the electrical network (grid) of the country in question. Table 3.3 shows the installed wind capacity of the main

Fig. 3.16 A wind turbine in Co. Offaly, Ireland; its expected output approaches 10 GWh per year. *(Photo: J.R.H. Ross.)*

68 Sustainable energy

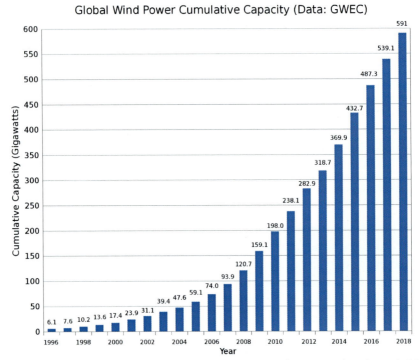

Fig. 3.17 Global wind power cumulative capacity. *(From https://en.wikipedia.org/wiki/Wind_power_by_country)*

Table 3.3 Installed wind power capacities in the major producing countries.

Country	2006	2019	% of national electricity production (2019)
China	2.6	236.4	5.4
United States	11.6	105.5	6.9
Germany	20.6	61.4	20.6
India	6.3	37.5	4.1
Spain	11.6	25.8	20.4
United Kingdom	2.0	23.5	19.8
France	1.6	16.6	6.2
Brazil	0.24	15.4	8.9
Canada	1.5	13.4	5.2
Italy	2.1	10.5	7.1
Denmark	3.1	6.1	53.0
Netherlands	1.6	4.6	9.5
Ireland	0.75	4.2	31.6
Belgium	0.19	3.9	10.2

Major producers of wind energy arranged in order of total capacity installed in 2019.
Data from Wikipedia and ourworldindata.org.

current users of wind power in 2019 as compared with the capacities of these countries in 2006. China is well ahead of other countries, having increased its capacity from only 2.6 GW in 2006 to almost 100 times as much, 236 GW, in 2019. Another country where there has been a significant increase in the use of wind energy is Brazil for which the comparable figures are 0.24 and 15.4 GW. An example of a country which now produces more than half of its electricity from wind is Denmark while another country with a high proportion of wind energy is Ireland for which the proportion is 31%.

There are many advantages of wind power over other technologies but there are also several disadvantages. The main advantage is that the installation and subsequent maintenance costs for wind turbines are relatively low and that the electricity can often be generated close to where it is needed, thereby reducing transmission losses. The main disadvantage is that wind power is very variable and there is therefore a need to have alternative standby generation capacity for periods when demand is high but the wind velocity is too low to provide significant electricity. Another problem is that turbines are relatively noisy and so they cannot be sited too close to populated areas. Wind turbines can also present visual problems, although opinions are divided on this topic, wind farms being liked by some and abhorred by others. One solution to the problem of visibility is to install them in the sea, some distance off-shore, where the winds can also be more powerful; however, the installation and maintenance costs then become more significant. Environmental problems also exist; for example, the installation of a wind farm at Derrybrien in the West of Ireland in 2003 caused dangerous land-slides with the result that $450,000\,\mathrm{m}^3$ of peat descended on the valley below, causing much environmental damage.

Solar power

As shown schematically in Fig. 3.18, when light hits a solid, electrons are emitted from the surface if the wavelength of the incident radiation is sufficient to detach the electrons from the material that is radiated. (In Fig. 3.18, the material radiated is shown as a metal but it could equally have been a semiconductor.) Photoemission has many applications but the use of relevance to the present discussion is in the operation of photovoltaic (PV) devices used to generate electricity when exposed to daylight. In a photovoltaic device, the electron, rather than being ejected into the surrounding space, is excited to a higher energy level and can be collected by a suitable collector. The first example of a solar cell operating on this principle was

70   Sustainable energy

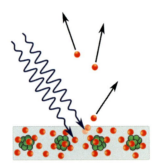

Fig. 3.18 Schematic depiction of photoemission. The emission of electrons from a metal plate caused by light quanta. *(From https://en.wikipedia.org/wiki/Photoelectric_effect)*

demonstrated by the American inventor, Charles Fritts, in 1883; his system, which exhibited a very poor efficiency, consisted of a layer of selenium covered by a thin conducting layer of gold and the first rooftop solar array using this combination was installed in New York as early as 1884.[h]

Solar cells are now generally made using crystalline silicon and in modern units these are usually in the form of thin films. The construction of a single cell is similar to that of a solid-state diode and consists of a thin P-type Si layer on top of an N-type Si layer. These are sandwiched between metal plates, the top layer having an open structure to allow light through. This structure is then coated by a thin glass window. A series of cells are then mounted together to then form a module (see Fig. 3.19), these being arranged either in series or parallel, depending on the desired output voltage. A number of these modules are then combined to form a solar panel and then a complete PV system. (As the panels will also gradually heat up, the removal of heat from the system also has to be taken into account.) As a PV system produces direct current, the output is fed to an inverter to convert it to AC for use in normal electrical systems or to feed it to the electrical grid. For domestic use of the electricity, it may be necessary to store the electricity in batteries for later use since power is produced only during daylight hours (see Fig. 3.20). The great advantage of solar power compared with wind energy is that it is much more consistent and predictable, not being dependent on direct

---

[h] It is interesting to note that such early cells, although consisting of expensive materials and having low efficiencies, found use as exposure-timing devices in more complex photographic cameras up until the 1960s.

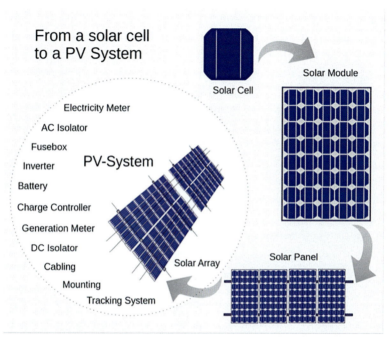

Fig. 3.19 From a solar cell to a PV system. *(From https://en.wikipedia.org/wiki/Solar_cell)*

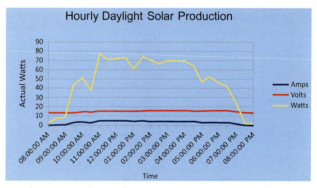

Fig. 3.20 Typical power production from a nominal 100 W solar module during daylight hours in August. *(From https://en.wikipedia.org/wiki/Solar_panel)*

sunlight but operating also on cloudy days, albeit with reduced efficiency on such days.

As a single PV cell can only be excited by only a limited range of wavelength of the incident light, a module may contain a variety of cells each accepting light of different wavelengths. The efficiency of an array can also be improved by focussing the light source in a so-called 'concentrator photovoltaic cell' (CPV). Other semiconductor materials are also used in modern devices, the most common of these being cadmium telluride and gallium arsenide. Many different formats have also been examined, ranging from solid materials to flexible films with suitable coatings. The efficiency of solar cells (defined as the percentage of incident radiation converted to electricity) is generally about 20% but this value is steadily being improved and higher values can be obtained for a given cell composition using a concentrator cell design. Fig. 3.21 shows that there has been a gradual improvement of module efficiencies of the various different types as a function of time. The cost of such modules are also gradually decreasing, as shown very schematically in Fig. 3.21.

A typical photovoltaic power plant is shown in Fig. 3.22. Such large-scale photovoltaic power plants (or colloquially 'solar farms') are becoming more and more common and one reads daily of new licences being granted for their construction. A number of governments worldwide have policies favouring such installations. However, as with wind energy, the rate at which installation can be achieved depends very much on the local capacities of the grid system in that country. There is no doubt that solar power will play an increasing role in decreasing universal greenhouse gas emissions.[i]

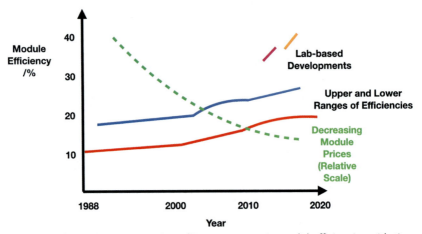

Fig. 3.21 Schematic representation of improvements in module fficiencies with time, showing upper and lower limits only. For fuller details, see https://en.wikipedia.org/wiki/Solar_cell_efficiency *(Redrawn from https://en.wikipedia.org/wiki/Solar_cell.)*

[i] The interested reader can obtain a free daily update of developments in the worldwide use of PV technology by subscribing to an electronic copy of PV Magazine (https://www.pv-magazine.com).

Less conventional energy sources 73

Fig. 3.22 Mount Komekura Photovoltaic power plant, January 2012. *(From https://en.wikipedia.org/wiki/Komekurayama_Solar_Power_Plant.)*

Solar power is also used for heating purposes using either reflective solar concentrators or via water heating. We will return in Chapter 4 to a discussion of the use of solar furnaces for hydrogen production and so will here only briefly discuss solar water heating systems. Fig. 3.23 shows a typical roof-mounted solar heating system used for domestic water heating. This consists of a series of tubes containing a high-temperature boiling fluid (in this case, a glycol/water solution) arranged to give optimum exposure to sunlight. The tubes absorb solar radiation, heating them to relatively high temperatures. The heated fluid in the tubes is then circulated through a specially constructed domestic hot water tank where heat is transferred to the water. Such a system works at all times of the year but has a much higher effect in the summer months. However, unlike solar PV panels, solar heating requires direct sunlight. On the day when the photograph was taken (just after the Autumn equinox), the roof fluid temperature rose to 55°C and the temperature at the base of the domestic tank was 49°C; in the summer months, these values can be much higher and the system in question is regulated to stop circulation if the hot water tank temperature exceeds 80°C. Electrical energy is required for the circulation of the fluid but the

Fig. 3.23 A solar water-heating installation. *(Photo by J.R.H. Ross.)*

circulation pump only operates if the temperature on the roof exceeds that in the storage tank; hence, assuming that the water would otherwise have been heated by electricity or by the combustion of fossil fuel, the savings in the emission of CO_2 associated with solar water heating can be quite significant. It is claimed that such systems can give a saving of up to 70% of the energy needed to provide a household's hot water requirements.

A recent development is the creation of dual-purpose modules which not only contain PV cells but also water-heating elements. As an example of such systems, the Italian company Greenetica Distribution has announced plans to build and sell a parabolic trough concentrating photovoltaic-thermal (CPVT) system. In this system, four parallel parabolic mirrors concentrate the solar radiation on two linear photovoltaic-thermal modules, the total width being 1.2 m. The photocells are based on InGaP, GaAs and Ge and work best at temperatures about 80°C. The parabolic mirrors give a concentration ratio of nearly 130 and approximately 91% of the incident sunlight is converted to heat or electricity, with an optimum output of 1 kW of electricity and 2.5 kW of thermal energy.[j]

---

[j] E. Bellini, PV Magazine, 6th October 2020.

Concluding remarks

It is clear that there have recently been some very significant advances in the use of renewable resources for the generation of heat and electricity and that many countries now produce significant amounts of renewable energy. However, as can be seen from Fig. 3.24, which shows the current level of renewable energy production in the European states, some countries are much more advanced than others. While some EU member states, notably Sweden, Finland and Denmark, exceeded their targets (these being shown as black lines for each country), a number of states, particularly those on the right-hand side, have failed to do so and are well below the EU average of just below 20% renewable energy production.

It is to be expected that there will be a very significant increase in the use of some of the methods that have been discussed in this chapter over the next few decades, this as a result of a strong worldwide pressure to overcome the problem of global warming and to approach net-zero greenhouse gas emissions. The following chapter, which describes the production and uses of hydrogen, includes a discussion of some of the ways in which renewable energy can be used to improve the emissions from some important chemical processes involving the use of hydrogen.

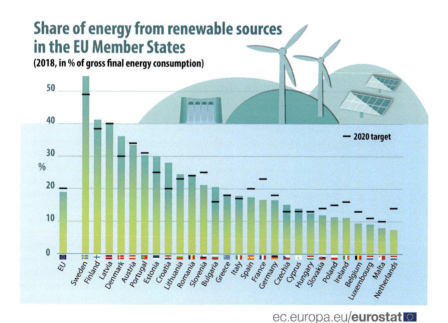

Fig. 3.24 The use of renewable energy in the EU. The black bars for each country denote their stated targets for 2020. *(From https://ec.europa.eu/eurostat/cache/infographs/energy/bloc-4c.html)*

CHAPTER 4

The production and uses of hydrogen

Introduction

Much has been written in both the technical literature and the popular press about a 'hydrogen economy' and the use of hydrogen as an energy carrier. We have all heard, for example, of the introduction of fleets of hydrogen-fuelled buses. Hydrogen as a fuel has the great advantage that only water is formed when it is combusted. However, as with many of the possible scenarios that are being considered as ways to reduce greenhouse emissions, there are a number of problems associated with the widespread introduction of hydrogen as a fuel. These include not only the fact that the most economical method of hydrogen production (steam reforming of methane) involves the production of large quantities of CO_2 but also that hydrogen transportation and storage both present problems. This chapter first discusses the production of 'grey' hydrogen by steam reforming, the currently most economical route, and also the use of CO_2 capture and storage (CCS) to produce 'blue hydrogen'. Also included is a discussion of the related processes used for the production of ammonia, methanol and hydrocarbon fuels, all of these currently being based on natural gas reforming. The current status of the production of hydrogen by electrolysis will then be discussed: if renewable electricity is used to provide so-called 'green hydrogen' using an electrolysis system, the derived products can also become 'green'. Hydrogen also has an important potential use as a fuel, in either direct combustion applications or in fuel cells, particularly in transport applications. The chapter, therefore, concludes with sections on transporting and storing hydrogen and on its use as a fuel in transportation applications.

The production of hydrogen from natural gas by steam reforming

In addition to its use as a fuel as discussed in Chapter 2, methane is currently the predominant source of hydrogen for industrial use. The majority of the

---

Sustainable Energy
https://doi.org/10.1016/B978-0-12-823375-7.00002-0

Copyright © 2022 Elsevier B.V.
All rights reserved.

77

78 Sustainable energy

hydrogen in general use is produced by the well-established steam reforming route although some is now produced by 'autoreforming', a topic to be discussed further below. The annual global production of hydrogen is about 70 million tonnes, of which approximately three quarters is formed from natural gas and this route uses approximately 6% of global natural gas production. (See Box 4.1 for a classification of the routes by which hydrogen is produced; some hydrogen is still made by coal gasification, as described in Chapter 2.) An alternative route for hydrogen production is the electrolysis of water (also discussed further below and in more detail in Chapters 7 and 8) but the cost of the hydrogen so produced is currently too high to make this method of production competitive except when it is used in very specialised situations. The data in Fig. 4.1 show that there is a continuously increasing use of pure hydrogen. Approximately 50% is used for refinery purposes (light blue bars), and much of the remainder (dark blue) is used for ammonia production; the small amount remaining ('other', ca. 5%) finds a variety of uses, including as a fuel for space-craft. The data in Fig. 4.1 do not include the hydrogen that is produced as a component of *syn*-gas (a mixture of H_2 and CO). In 2018, this additional quantity amounted to about 42 million tonnes of which about 12 million tonnes were used for methanol production, about 4 million tonnes were used for direct reduction of iron in steel production and the remaining 28 million tonnes was used for fuel

BOX 4.1 Colour-coded nomenclatures for hydrogen production

In some of the literature on hydrogen production, hydrogen is listed under a set of colour-coded headings:

Black hydrogen—produced by gasification of coal

Brown hydrogen—produced by gasification of lignite

Grey hydrogen—produced by steam reforming of natural gas without CCS

Blue hydrogen—produced by steam reforming of natural gas with CCS

Green hydrogen—produced by electrolysis using renewable electricity

The most commonly encountered of these are *grey*, *blue* and *green* hydrogen since coal and lignite gasification are gradually being phased out. Nevertheless, *black* hydrogen will still be produced for many years to come in countries such as China and India that have low reserves of natural gas.

The term CCS ('carbon capture and storage') is sometimes replaced by the term CCUS ('carbon capture, use or storage') in situations in which any CO_2 formed is used in another related process rather than being stored.

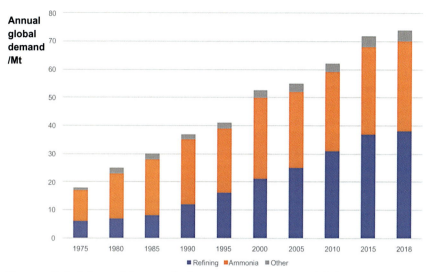

Fig. 4.1 Global demand for pure hydrogen since 1975. *(IEA, Global demand for pure hydrogen, 1975–2018, IEA, Paris https://www.iea.org/data-and-statistics/charts/global-demand-for-pure-hydrogen-1975-2018.)*

and feedstock purposes (e.g. for Fischer Tropsch synthesis).[a] Some of the methods currently available for hydrogen production are now described and subsequent sections describe some other methods of production as well as some of the current uses of hydrogen, these including the synthesis of ammonia, methanol and synthetic fuels (See Box 4.2).

The arrangement of a typical plant of the type that is used for pure hydrogen production is shown schematically in Fig. 4.2.

The steam reforming reaction may be depicted as follows:

$$CH_4 + H_2O \rightleftharpoons CO + 3H_2 \quad \Delta H° = +206.2 \text{ kJ}$$

This endothermic reversible reaction is shown as being reversible as the conversion only approaches 100% at the very high temperatures required for the process, 850–900°C.[b] The reaction is in most applications catalysed by a Ni-containing material and the reaction is operated in the presence of excess steam to prevent the deposition of carbon by either the methane decomposition reaction:

---

[a] Data from 'The Future of Hydrogen: Seizing Today's Opportunities', IEA Report prepared for the G20 meeting in Japan in 2019 (https://www.iea.org/topics/hydrogen/).
[b] Because the hydrogen produced in the reaction is generally required to be at high pressures, the steam reforming reaction is also carried out at high pressures, typically above 25 atm. The consequence is that the temperature needed for the conversion desired has to be increased further.

BOX 4.2 Carbon capture and storage (CCS)

As discussed in the main text, the CO_2 emissions from many different industrial processes can be very high. In many cases, such as in ammonia synthesis, CO_2 is removed from the product gases resulting from steam reforming of natural gas by well-established amine scrubbing processes; some of the CO_2 is then used in a variety of downstream processes such as in methanol synthesis, even though large proportions are still emitted to the atmosphere. In other processes such as electricity generation, the CO_2 is currently in almost all cases emitted to the atmosphere. In order to improve these situations, methods of storing the CO_2 are being developed, these mostly involving storage in deep subsurface geological formations such as spent oil or gas fields or disused salt or coal mines. Most of this gas that is trapped will be stored indefinitely but some may also be re-used, for example for enhanced recovery of oil, coal-bed methane and gas. The requirements for successful storage include that they should have a rock structure to prevent the escape of CO_2, preferably at a depth of at least 800 m, and that the chambers should be large enough to store the required amount of CO_2 for the lifetime of the site. CCS using such storage facilities present many problems, including the fact that the CO_2-rich gases to be stored must be transported to the storage site. This will normally require the use of suitable pipelines; although many such pipelines already exist, it has been estimated that an additional 70,000 to 120,000 km of the pipeline would be required globally by 2030 if CCS was to be widely introduced. Any storage facility would also need to be monitored continuously for leakage and for any environmental effects throughout its lifetime. Perhaps one of the most important factors is that there is worldwide only a relatively small number of suitable storage sites available and so there is a finite limit to the amount of storage that can be achieved. One of the consequences of this is that there will be a limit to the amount of 'blue hydrogen' that can be produced.

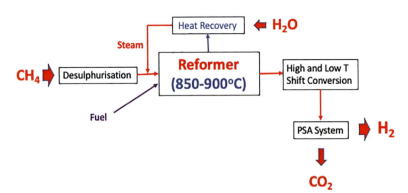

Fig. 4.2 Schematic representation of a plant for the steam reforming of methane to produce pure hydrogen.

$$CH_4 \rightleftharpoons C + 2\,H_2 \quad \Delta H^\circ = 74.9\,\text{kJ}$$

(a reaction that predominates at the high temperatures of the steam reforming process) or the Boudouard reaction:

$$2\,CO \rightleftharpoons C + CO_2 \quad \Delta H^\circ = -172.4\,\text{kJ}$$

(a reaction that predominates at lower temperatures such as those at the exit of a steam reformer system). At the high temperatures required for steam reforming, CO is the main additional product formed but the reactor effluent can also contain small quantities of CO_2, this being formed in the water-gas shift equilibrium reaction:

$$CO + H_2O \rightleftharpoons CO_2 + H_2 \quad \Delta H^\circ = -41.2\,\text{kJ}$$

When hydrogen is the main desired product, for example for use in ammonia synthesis, the product gas from the steam reformer is then passed through two water-gas shift reactors operating at lower temperatures; these convert the majority of the CO into CO_2 which is then removed in a pressure-swing absorption (PSA) system. (Any remaining traces of CO, a poison for ammonia synthesis catalysts, are then removed by methanation, the reverse of the steam reforming reaction.) Box 4.3 gives some more detail on the catalytic processes occurring in the steam reforming reaction.

Fig. 4.4 shows the exterior of a typical steam reforming plant. The scale of this plant can be seen from the size of the reactor housing. This housing contains approximately 300 parallel reactor tubes, each typically $10\,\text{m}$ in length, that are heated by the combustion of natural gas in burners spaced along the tubes. As discussed in Box 4.3, the reaction rate is determined by the rate at which the reactants reach the active catalyst surface. Hence, the process can easily be scaled down (although further scaling up is not economically justifiable due to reactor costs) so that it is possible to produce hydrogen by steam reforming in relatively small plants compared with ones such as that shown in Fig. 4.4. Hence, for example, it is possible to envisage a situation in which hydrogen to be used for transportation purposes could be generated at the fuelling station rather than having to be transported to that station using sophisticated high-pressure transporters. Fig. 4.5 shows a small-scale steam reforming plant designed for this purpose. This plant contains all of the elements of the full-scale plant shown in Fig. 4.4 (hydrodesulphurisation, water-gas shift, pressure swing adsorption) but the reactor tubes are now only approximately $2\,\text{m}$ in length. HyGear, the company making this unit, says that their plants of various different sizes produce hydrogen up to 99.9999% purity at a flow rate of up to $500\,\text{Nm}^3/\text{h}$ and at pressures

BOX 4.3 Catalysis and the steam reforming of methane

What is a catalyst? A catalyst is a material which, when added to a chemical reaction mixture, increases the rate of the chemical reaction but is itself not used up in the reaction.[c] In the case of the steam reforming of methane, the reaction would require some extremely high temperatures in order to proceed at an acceptable rate if no catalyst was present. Using a catalyst, the steam reforming reaction occurs at much lower temperatures, at ca. 450°C and above, by adsorbing the reactants (and products) on the surface of active material, in this case generally metallic nickel, in such a way that the all-over activation energy for the process is significantly reduced. The attainment of good conversions in the steam reforming reaction is however further complicated by the fact that the reaction is very endothermic, this having the consequence that high temperatures are necessary in order to obtain the desired conversions, particularly as the pressures are also relatively high (see main text).

The steps occurring during the catalytic conversion of gaseous methane and water (the feed) to CO and hydrogen (the products) are shown very schematically in Fig. 4.3. The feed enters the reactor, in this case a long tubular packed-bed (or more frequently, a large assembly of such tubes in parallel, as discussed in the main text) containing a suitable catalyst. The catalyst generally comprises of nickel on a temperature-resistant support material such as α-alumina or magnesium aluminate spinel in the form of Raschig rings (or even more complicated shapes). Such shapes are used to optimise gas flow through the reactor and to enable the reaction mixture to reach all parts of the external surfaces of the support materials. These generally have relatively large pores and the feed molecules diffuse to the nickel surfaces within the support through these pores. Adsorption and dissociation of the methane and water then occur and reaction occurs between these species on the Ni surface. Finally, the products (carbon monoxide and hydrogen) desorb from the nickel and diffuse out of the pores and hence towards the exit of the reactor. The steam reforming reaction is one of the very few catalytic reactions in which the rates of the individual chemical surface processes occurring do not determine the rate of the reaction; instead, the conversion achieved in the tubular reactor depends solely on the rates of external and internal diffusion, the majority of the reaction occurring near the entrances to the pores where equilibrium conversions are rapidly achieved. Everything that occurs in the reactor is therefore determined by the thermodynamics of the various reactions taking place and the kinetics of the reactions involved have little or no influence. It is for this reason that the reaction must be operated using excess steam in order to avoid conditions that would thermodynamically permit carbon deposition to occur in any position in the reactor.[d]

Many of the transition metals of Group 8 will bring about the steam reforming reaction. However, of these, Ni is the metal most commonly used, mainly because it

Continued

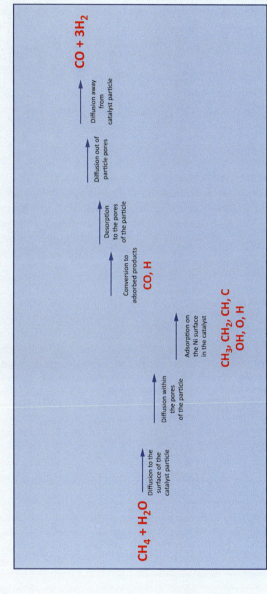

Fig. 4.3 Schematic representation of the stepwise conversion of natural gas to hydrogen in the steam reforming reaction. Note that the reaction will occur in the reverse direction (methanation of CO) if the temperature of the reaction is much lower. The possible existence of the water-gas shift reaction (giving CO_2) is not included in this scheme as the equilibrium proportion of CO_2 under these reaction conditions is negligible. Note that species shown below the level of the feed CH_4 and H_2O are formed exothermically (release energy), while those above this level are formed endothermically (require energy).

Continued

84   Sustainable energy

BOX 4.3 Catalysis and the steam reforming of methane—cont'd

is much cheaper and more abundant than the other Group 8 metals. It is also important in the choice of active component that the active metal is not oxidised by the water present under reaction conditions according to the reversible reaction:

$$M + H_2O \rightleftharpoons MO + H_2$$

For a nickel, the metal remains in its reduced form as long as the ratio P_{H2}/P_{H2O} is greater than about 3×10^{-3}. Cobalt will also remain in its reduced state as long as this ratio is greater than 2×10^{-2}. However, iron requires significantly higher hydrogen partial pressures to ensure that the metal remains reduced; if the oxidised state considered is FeO, the ratio has to be greater than 5.0. For the noble metals, the potential oxidation of the metal is not an important limitation as they all remain in the reduced form under these conditions. It has been found that they all carry out the reforming reaction very efficiently. However, the costs of the noble metals are too high for them to be used in practice. Rostrup Nielsen has reported that the order of activities of the noble metals for the steam reforming of ethane is: Rh, Ru > Ni, Pd, Pt > Re.[e]

[c] A full discussion of heterogeneous catalysis and of the preparation, characterisation and application of catalysts is beyond the scope of this book. A reader who does not have a basic knowledge of these topics is advised to refer to a textbook such as 'Contemporary Catalysis - Fundamentals and Current Applications' Julian R.H. Ross, Elsevier, 2019; ISBN: 978-0-444-634740-0.

[d] The present author has reported work on the steam reforming of methane carried out at low pressures (ca. 0.03 atm.) in the temperature range 500–680°C when CO_2 was also a possible product. The work showed that CO was the primary product of the reaction and that the adsorption of methane on the Ni surface was the rate determining step under these conditions. (see J.R.H. Ross and M.C.F. Steel, 'Mechanism of the Steam Reforming of Methane over a Coprecipitated Nickel-Alumina Catalysts', J. Chem. Soc., Faraday Trans. I, 69 (1973) 10–20.)

[e] J.R. Rostrup Nielsen, J. Catal., 31 (1976) 110.

up to 300 bar (g). The system illustrated is a HyGEN 150 model that is built in a 40 ft. container. The hydrogen production rate is described as being in the range 123 to 141 Nm3/h, depending on the purity required (99.5 to 99.999999%), at a pressure from 1.5 bar(g) to 7 bar(g).

It is possible to produce hydrogen on an even smaller scale using microreactor technology. As discussed in Box 4.4, such microreactors contain thin layers of catalyst confined in microchannels in close proximity to the heating elements used to provide the necessary energy for the reaction. Such reactors give much higher catalyst effectiveness than the more conventional system described above by minimising diffusion limitations. Many different designs of such microreactors have been described in the literature but there is as yet no indication of their commercial application.

Fig. 4.4 A typical large-scale plant for the steam reforming of natural gas to produce pure hydrogen This is a plant for the production of ammonia at a rate of 1500 MTPD; the secondary reformer is also visible. Photograph reproduced with kind permission of Haldor Topsoe. (See footnote 't'). *(Photograph reproduced with permission from Haldor Topsoe's White Paper on the New SynCOR Ammonia process; www.topsoe.com.)*

Fig. 4.5 A small-scale steam reforming unit for the local production of hydrogen at hydrogen-refueling stations. *(Photograph kindly supplied by HyGear Ltd.)*

BOX 4.4 Microreactors and their use for hydrogen production

There has been a great deal of work on the use of microreactor systems for reactions such as the steam reforming of methane. An important aspect of a microreactor is that heat exchange to and from the active catalyst is easily achieved as the catalyst is generally situated very close to the wall of the channel in which it is situated and heat is easily transmitted in either direction at that wall, reducing heat transfer limitations. The topic of microreactors has been well reviewed by a number of groups active in the field. A particularly useful review is one by Kolb and Hessel in which a description is given of some of the various reactor structures that had been reported in the literature up until 2004.[f] Fig. 4.6 is a diagram taken from the review that shows the structure of a typical microreactor used as part of a complete system for the steam reforming of methanol, showing typical dimensions of such a reactor. The length of each reaction channel is only several cm. In this example, a hydrogen-rich gas containing traces of CO produced by the steam reforming of methanol enters on the right-hand side and passes through the appropriate heat exchangers as well as the central section containing a selective oxidation catalyst before passing to a fuel cell. The important aspect of this type of reactor is that heat and mass transfer limitations are minimised, vastly improving the efficiency of the process involved, and that the quantity of catalyst required is also therefore significantly reduced.

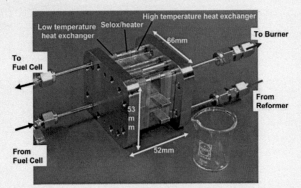

Fig. 4.6 Typical dimensions of a microreactor An example of the structure of a microreactor designed for the selective oxidation of traces of CO from a hydrogen stream emerging from an ethanol steam reforming reactor prior to its admission to a fuel-cell system. *(From 'Micro-structured Reactors for Gas Phase Reactions'. (2004). Chem. Eng. J., 98, 1–38. Reproduced with the kind permission of Elsevier.)*

[f] G. Kolb and V. Hessel, 'Micro-structured Reactors for Gas Phase Reactions', Chem. Eng. J. 98 (2004) 1–38.

The energy required to control the temperature of the catalyst assembly in a microreactor used for steam reforming of methane can be applied by electrical heaters. However, an alternative is to insert between each steam reforming element a second catalyst layer in which methane combustion takes place. Fig. 4.7 shows such a possible structure as described in a paper by Junjie Chen and colleagues that concerns the modelling of a thermally integrated microreactor system in which a very active rhodium catalyst, used for steam reforming, is contained in one set of channels and an equally active Pt catalyst, used for methane combustion, is contained in the others, there being good thermal conductivity between the two sets of channels.[g] (The authors stress the importance of the presence of highly active catalysts; in such a micro-reactor construction, the surface reaction becomes the important step and diffusion limitations are negligible.) Fig. 4.8 shows the results of modelling the temperature profile through a single reforming channel and the corresponding methane mole fractions, the methane conversions and the fluid velocities through the same channel. The conversions of methane are greatest near the walls of the channel and the highest conversion is attained at the end of the channel (The rate of flow is also shown, this being greatest at the end of the channel as a result of the increased number of molecules in the product gases compared with the reactants.)

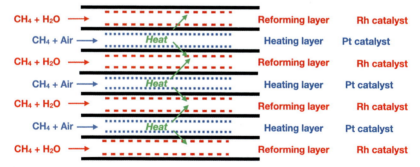

Fig. 4.7 A microreactor system containing parallel steam reforming and methane oxidation channels used for modelling the reaction; there is good heat transfer between the channels. *(Derived from 'Compact Steam-Methane Reforming for the Production of Hydrogen in Continuous Flow Microreactor Systems.' (2019). A.C.S Omega, 4, 15600–15614.)*

---

[g] J.J. Chen, W.Y. Song and D.G. Xu, 'Compact Steam-Methane Reforming for the Production of Hydrogen in Continuous Flow Microreactor Systems', ACS Omega, 4 (2019) 15600–15,614.

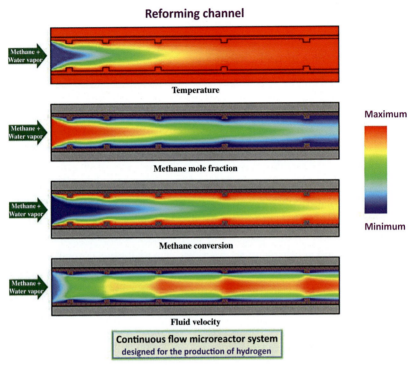

Fig. 4.8 The results of modelling the temperature profile through a single reforming channel and the corresponding methane mole fractions, the methane conversions and the fluid velocities through the same channel. *(From 'Compact Steam-Methane Reforming for the Production of Hydrogen in Continuous Flow Microreactor Systems.' (2019). A.C.S Omega, 4, 15600–15614, with kind permission of American Chemical Society.)*

A significant addition to the cost of operating steam reforming plants of the type described above is related to the cost of the removal of CO_2 from the effluent of the reactor. One method of avoiding this problem is the use of a reactor system incorporating a membrane separation system of the type shown schematically in Fig. 4.9. It is well established that membranes of palladium or its alloys (e.g. Pd/Ag) can be used to separate out the hydrogen formed. The hydrogen is dissociatively *adsorbed* on the Pd surface before being *absorbed* in the bulk of the Pd; the absorbed atomic hydrogen then diffuses through the bulk of the palladium to the other side of the membrane where it is desorbed as molecular hydrogen. The driving force for this diffusion is the hydrogen concentration gradient through the membrane.

There has been significant effort in designing and testing systems that allow membrane separation to occur within the reformer system. Fig. 4.10, taken from a review by Nikolaidis and Poullikkas, shows one such

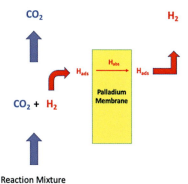

Fig. 4.9 Schematic representation of a process for hydrogen separation from a steam reforming product stream using a membrane involving a metal such as Pd or a Pd/Ag alloy.

system.[h] In this design, the catalyst particles are placed in the spaces between the membrane tubes. As hydrogen is withdrawn from the product mixture, the equilibrium of the steam reforming reaction is driven to the right, allowing high equilibrium conversions to be achieved at lower temperatures than in conventional systems.

This approach is illustrated schematically in Fig. 4.11. The effectiveness of the combination of catalyst and membrane will depend on both the catalytic activity attained as well as the permeability of the membrane under the reaction conditions. Such a system will thus require careful control of the reaction conditions used, particularly the temperature of both the catalyst and membrane.

The use of other metals for membrane systems is discussed briefly in Box 4.5.

The production of hydrogen from natural gas by other methods

Syngas can be produced from methane by two other reactions, partial oxidation, often now referred to as 'autoreforming':

$$CH_4 + 0.5\,O_2 \rightarrow CO + 2\,H_2$$

and CO_2 reforming, better known as 'dry reforming':

$$CH_4 + CO_2 \rightleftharpoons 2\,CO + 2\,H_2$$

These two processes will now be discussed briefly in turn.

---

[h] P. Nikolaidis and A. Poullikkas, 'A Comparative Overview of Hydrogen Production Processes', Renewable and Sustainable Energy reviews, 67 (2017) 597–611.

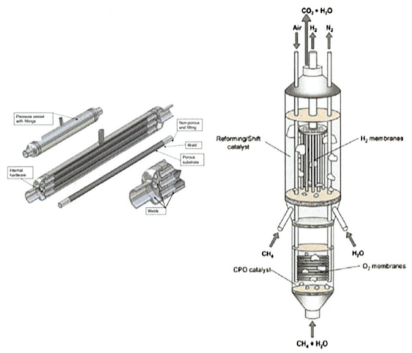

Fig. 4.10 A membrane reactor containing multiple Pd membrane tubes designed for the steam reforming of methane. *(From 'A Comparative Overview of Hydrogen Production Processes.' (2017). Renewable and Sustainable Energy Reviews, 67, 597–611. Reproduced with kind permission of Springer.)*

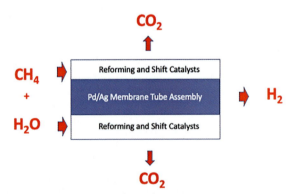

Fig. 4.11 Schematic representation of the combination of a methane steam reforming catalyst and a Pd-based membrane for simultaneous selective hydrogen removal.

BOX 4.5 The possible use of other metals such as tantalum as a membrane material

The composition of the membranes used for hydrogen separation has also received significant attention. For example, Rothenberger et al. have reported on the hydrogen permeability of tantalum-based membrane materials; they found that the rate of permeation was low, probably as a result of surface contamination.[i] In this context, it is probably worth noting some of the work from the PhD thesis work of the present author.[j] As part of a study of the adsorption of hydrogen sulphide and hydrocarbons such as neopentane on evaporated metal films, he showed that the uptake of hydrogen in clean films of tantalum evaporated under high vacuum conditions was facile and that the quantities absorbed increased with temperature in the range studied, 70 to 132°C; the uptakes corresponded to up to 0.14H per Ta atom in both the evaporated film and the filament from which the film had been evaporated, this being maintained at the same temperature. (As part of this work, it was also shown that uptakes of hydrogen in Pd of up to 0.6 H/Pd could be obtained in prolonged experiments but the total uptake possible for Ta was not measured.) The work showed the importance of having a clean metal surface to dissociate the hydrogen[k]: the effect on hydrogen absorption of adsorbing sulphur from the dissociation of H_2S was examined and it was found that a monolayer of S species completely stopped hydrogen absorption with both metals. However, desorption of hydrogen from one or other bulk hydride through the S-layer was still possible, this indicating that hydrogen atoms could still transfer through the contaminating S layer. Further, it was shown that it was possible to have continued absorption of hydrogen from the gas phase if the hydrogen gas was dissociated into atoms at the Ta filament from which the film had been evaporated when this filament was heated in the presence of hydrogen.

[i] K.S. Rothenberger et al., 'Evaluation of Tantalum-based Materials for Hydrogen Separation at Elevated temperatures and Pressures', J. Membrane Sci. 218 (2003) 19–37.
[j] J.R.H. Ross, PhD Thesis, Queen's University of Belfast, 1966; see also M.W. Roberts and J.R.H. Ross, 'The Interaction of Hydrogen Sulfide and Hydrogen with Palladium and Tantalum Films', in Reactivity of Solids, Edited by J.W. Mitchell et al., John Wiley and Sons, N.Y., 1969.
[k] Until the 1960s, work on the Pd/H and Ta/H systems had used electrochemical methods to incorporate the hydrogen, under which conditions possible surface contamination was less critical.

Autoreforming

The partial oxidation reaction is shown above as being irreversible as 100% conversion is thermodynamically possible at the higher reaction temperatures used (typically up to 1000°C) and the product has a syngas ratio of 1:2. No carbon deposition is allowed thermodynamically under normal reaction temperatures.[1] If pure hydrogen is required as the product gas, steam is added to the reaction mixture and this is available downstream to strip out the CO formed by carrying out the water-gas shift conversion as is shown in Fig. 4.12. In this case, a total of three hydrogen molecules are formed from the reaction of one molecule of methane. Many major plant construction companies are now installing autoreforming units in place of conventional steam reforming plants for hydrogen production. For example, Haldor Topsøe has published very informative 'White Papers' on the use of autoreforming in ammonia and methanol manufacture.[m] Both of these processes will be discussed further below in relation to some of the uses of hydrogen. Suffice it to say here that the air separation unit supplies a ready supply of pure nitrogen for the ammonia synthesis process while the CO_2 formed in the water-gas shift reactor is an excellent feed material for the synthesis of methanol. (Omitting a water-gas shift system would mean that the effluent gas could be used directly for the more conventional methanol synthesis route using CO as a reactant). Another important point to be made here is that the total size of an autoreforming plant is small compared with a conventional

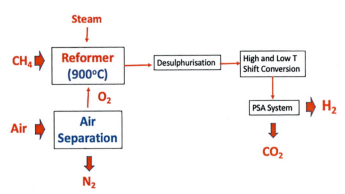

Fig. 4.12 Schematic representation of an autothermal reactor for the partial oxidation of methane to produce pure hydrogen.

---

[1] In practice, the methane entering the reactor is first oxidised exothermically to give water and CO_2; this is followed by the endothermic reactions, steam reforming and dry reforming, in the rest of the reactor.'

[m] 'New Syncor ammonia process'; 'Methanol for a more sustainable future - Electrified chemicals'; 'Lower your carbon footprint while boosting your CO output using Topsoe's unique ReShift technology'. Haldor Topsoe white papers, www.topsoe.com.

steam reforming plant. This has the important result that investment costs are lower. Additionally, the total capacity of the plants can also be much higher. We will return to this topic below when discussing the uses of hydrogen in ammonia and methanol synthesis.

Dry reforming of methane

As shown in the equation above, the dry reforming reaction produces a mixture of CO and H_2 in the ratio 1:1. Dry reforming is a topic that has received much research attention for a number of years but the reaction is not used in practice, partly because of problems with carbon deposition on the catalysts used and partly because there are few uses for the product gases in the ratio 1:1. The lack of such uses of the products in chemical processes precludes the potential use of the reaction for CO_2 mitigation, a factor generally overlooked by authors of research papers on the subject. However, a combination of steam reforming and dry reforming (i.e. co-feeding H_2O and CO_2) can be used as the product ratio is then much more usable. However, the problem of carbon deposition on the catalysts still persists.[n]

Methane pyrolysis

The technology for the production of hydrogen and syngas using the pyrolysis of coal has been known for many years. More recently, the method has been applied to the pyrolysis of hydrocarbon materials in which the hydrocarbon is the sole source of the hydrogen produced:

$$C_nH_m \rightarrow nC + m/2\ H_2$$

For low molecular weight hydrocarbon feeds (boiling point up to 200°C), hydrogen is produced directly. For higher molecular weights, the hydrocarbon is gasified in two steps; in the first, hydrogasification of the hydrocarbon feed to form methane is carried out:

$$C_nH_m + H_2 \rightarrow nCH_4$$

---

[n] The present author has reviewed some of the recent literature on dry reforming catalysts in an article entitled: 'Syngas production using carbon dioxide reforming: Fundamentals and perspectives'. (In 'Transformation and Utilization of Carbon Dioxide', Edited by B.M. Bhanage and M. Arai, pages 131–161, Springer, (2014).

94   Sustainable energy

and this is then followed by pyrolysis of the methane produced:

$$CH_4 \rightarrow C + 2H_2$$

It should be recognised that the hydrogen so produced can be considered as 'blue' (Box 4.1) since the carbon formed can easily be stored or alternatively can be used for soil remediation.

Electrolysis of water

The electrolysis of water to produce hydrogen is a well-established method of producing pure hydrogen:

$$2H_2O \rightarrow 2H_2 + O_2$$

As is discussed in greater detail in Chapter 7 the value of ΔG° for this reaction is positive and the decomposition of water would require very high temperatures for it to occur spontaneously; the reaction can only be brought about at and around room temperature by applying an electrical potential sufficient to exceed this positive value of ΔG°.[°] Fig. 4.13 gives a schematic representation of an electrolysis system used for the reaction. In order to safely segregate the hydrogen and oxygen formed in the electrolysis reaction, it is necessary to introduce some sort of barrier (a membrane) between the anode and cathode of the electrochemical reactor. Common systems used are proton exchange membrane (PEM) cells and solid oxide electrolysis (SOEC) cells. PEM cells operate by transferring hydrogen ions (formed at the anode) from the anode compartment by way of an ion-conducting polymer membrane to the cathode compartment where hydrogen gas is then liberated at the cathode. With the SOEC cells, hydroxyl species formed at the cathode are transformed to oxygen ions at the solid membrane surface and are then transferred through the membrane, finally reacting at the anode to form molecular oxygen. These systems therefore differ in the species that are transferred from one side of the cell to the other, hydrogen or oxygen. Further, as SOEC cells operate most effectively at elevated temperatures, part of the energy

---

[°] An electrolysis reaction occurs at equilibrium if Go $= -$ nFE, where n is the number of electrons involved in the process, F is the Faraday equivalent (the charge carried by one mole of electrons, 96,494 coulombs per mole) and E is the potential applied to the cell. In practice, a higher potential (an 'overpotential') has to be applied to cause the electrolysis reaction to proceed at a significant rate. (See Chapter 6 for a fuller discussion.)

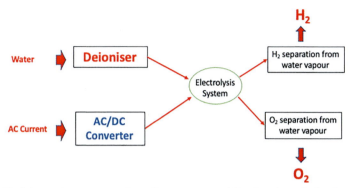

Fig. 4.13 Schematic representation of a system used for the electrolysis of water to produce both pure H$_2$ and pure O$_2$.

needed to drive the reaction must be supplied thermally rather than electrically.[p] In practice, an installation used to produce hydrogen requires water purification and an AC/DC converter to provide the necessary direct current voltage, as well as systems to dry the product gases (Fig. 4.13). The purities of both the hydrogen and oxygen produced are very high.

Electrolysis as a method for the production of hydrogen is currently seen only as a method to be used in very specialised applications since the cost of operation is high as a result of the high cost of electricity. The consequence is that only approximately 5% of current global hydrogen production is by electrolysis. However, the outlook is that we shall soon see a change in this scenario with the movement towards renewable electrical energy generation. If the costs of renewable electrical power can be reduced still further and the demands for local supplies of hydrogen increase, we are likely to see a steady move towards the production of 'green' hydrogen by electrolysis. This topic is handled further in Chapter 7.

---

[p] A detailed description of electrochemical processes is beyond the scope of this book. For further information, an interested reader should consult a modern textbook on Physical Chemistry or Electrochemistry.

96   Sustainable energy

The generation of hydrogen from biomass by various processes

As will be discussed in much greater detail in Chapter 5, the term *biomass* applies to a wide variety of materials, ranging from commonly available commodities such as grass and corn stover to timber and agricultural waste. Many of these materials contain bio-polymers that consist of cellulosic and hemicellulosic species as well as lignin and other molecules. All of these contain some level of hydrogen and a variety of methods, to be discussed later, have been used either to extract this hydrogen or to convert the biomass components into useful chemicals. As long as the biomass used is fully 'renewable', the products are generally considered as 'green', not contributing to greenhouse gas emissions when finally reaching the stage of disposal.[q] Other processes for biomass utilisation are also described in Chapter 5.

Comparison of hydrogen production costs for different processes

Nicolaides and Poullikka have assembled a series of useful estimates of the costs of hydrogen production by various different methods including those described above. Table 4.1 summarises some of their most relevant data. The interested reader should consult the original reference for details of the sources from which these data have been extracted. It can be seen that autoreforming consistently gives among the lowest costs. As these data were assembled during the period 1992 to 2007, it is probable that the costs of production of hydrogen using autoreforming and electricity from solar PV will now both be significantly lower due to improvements in autoreforming plant design and improved PV efficiencies. For an explanation of the term Carbon Capture and Storage (CCS), see Box 4.2 above.

Methanol production

We now discuss a number of important industrial processes that use the hydrogen and syngas produced by the various routes. Methanol synthesis makes use of a syngas mixture according to the following reaction:

$$CO + 2\,H_2 \rightarrow CH_3OH$$

The proportions of CO and hydrogen in the syngas are adjusted to the correct value by making use of the water-gas shift reaction, as described above. The catalyst used for the synthesis process is generally a Cu–

---

[q] Forestry operations are considered to be renewable if equivalent new plantation occurs following harvesting. However, destruction of forestry in the process of land clearance is to be abhorred.

The production and uses of hydrogen   97

Table 4.1 Hydrogen production costs using different methods.

Process	Energy source	Feedstock	Capital cost/ m$	H$_2$ cost/$ kg^{-1}
Steam reforming with CCS	Fossil fuels	Natural gas	226.4	2.27
Steam reforming without CCS	Fossil fuels	Natural gas	180.7	2.08
Coal gasification with CCS	Fossil fuels	Coal	545.6	1.63
Autoreforming with CCS	Fossil fuels	Natural gas	183.8	1.48
Methane pyrolysis	Internally generated steam	Natural gas	—[a]	1.59–1.70
Biomass pyrolysis	Internally generated steam	Woody biomass	53.4[b]	1.25–2.20
Biomass gasification	Internally generated steam	Woody biomass	149.3[c]	1.77–2.05
Solar PV and electrolysis	Solar	Water	12–54.5	5.89–6.03
Wind and electrolysis	Wind	Water	504.8–499.6[d]	5.89–6.03
Nuclear power and electrolysis	Nuclear	Water	—[a]	4.15–7.0

[a]No cost given.
[b]For a plant capacity of 72.9 tons per day.
[c]For a plant capacity of 139.7 tons per day.
[d]The higher cost presupposes the generation of electricity along with hydrogen while the lower cost is for hydrogen alone.
Data extracted from the reference of footnote 'g'.

$ZnO-Al_2O_3$ material similar to that used for the water-gas shift reaction. There are many large-scale methanol plants worldwide, a particularly high concentration of these being in Asia; the total global production of this important chemical was about 140 million tonnes in 2018 and it is predicted that that figure will double by 2030. There has recently also been an interest in producing methanol from CO_2; see Box 4.6.

Production of fuels using the Fischer Tropsch process

Another important use of syngas is in the Fischer-Tropsch (F.T.) process for the production of synthetic fuels. The reaction can be represented very roughly by the chemical equation:

$$nCO + (2n + 1)H_2 \rightarrow CnH_{2n+2} + nH_2O$$

98 Sustainable energy

BOX 4.6 Green methanol

There is currently significant interest in producing methanol by the hydrogenation of CO_2 rather than of CO. The process proceeds in two steps, the first of these being the reverse of the normal water-gas shift reaction discussed above:

$$CO_2 + H_2 \rightleftharpoons CO + H_2O$$

This is followed by the reaction of the CO with hydrogen to give methanol as described in the main text:

$$CO + 2H_2 \rightarrow CH_3OH$$

The all-over reaction is therefore:

$$CO_2 + 3H_2 \rightarrow CH_3OH + H_2O$$

The production of methanol from CO_2 therefore requires the use of more hydrogen than does its production from CO. The process described here could be used as a method of utilising CO_2 but it would only have any significant value if the hydrogen could be produced using renewable energy, for example by an electrolysis route using electricity produced using a renewable route such as one of those discussed in Chapter 2.[r] The methanol so produced has been termed 'green methanol'. See also Chapter 7 where an alternative method of using CO_2 for this reaction is discussed.

[r] Haldor Topsøe have recently described their work on the production of 'e-Methanol', otherwise called 'electrified methanol', in a brochure entitled: 'Methanol for a More Sustainable Future' (topsoe.com/emethanol). Their process uses a specially designed Cu-containing methanol catalyst 'MK-317 SUSTAIN' which they claim gives high selectivity for methanol synthesis from CO_2 and hydrogen, the latter being produced by electrolysis. Their process uses ca. 500 kWh of electricity to produce one ton of high-purity methanol. Topsøe's work on the production of hydrogen using electrolysis is discussed in Chapter 7.

The range of products obtained is quite large, ranging from methane to waxes, and the resultant mixture may also contain some oxygenated species. The actual product composition obtained in a particular plant is very dependent on the reaction conditions and the catalyst used.[s]

The F.T. process was initially developed in Germany during the 1930s for the production of fuels. The original process used coal as raw material for syngas production using Lurgi gasifiers. The process was then further developed in South Africa by Sasol during the years of Apartheid, this having been essential since South Africa was at that time solely dependent on domestic resources. Following the discovery of offshore gas fields in Mossel Bay

---

[s] For further information on this topic, the reader should refer to Chapter 12 of the book listed in footnote 'b'.

(400 km. east of Capetown) in 1969, the South African Petroleum, Oil and Gas Corporation (PetroSA) started to use syngas produced from natural gas as a feed for the FT process. This installation currently produces 15% of South Africa's transport fuel requirements.

The Fischer Tropsch Process is now also used a number of major 'Gas to Liquids' (GTL) plants in which the syngas required is produced from natural gas by partial oxidation:

$$CH_4 + 1/2\,O_2 \rightarrow CO + 2\,H_2$$

This reaction can be carried out using either air, when the product gas will contain nitrogen, or pure oxygen (Auroreforming), when a preliminary air separation step is required, as described above. Shell uses partial oxidation in several major commercial GTL plants; the first of this started operation in Bintulu, Malaysia in 1993 and another, which Shell claims is the world's largest, started operation in Qatar in 2011 ('Pearl GTL').

Production of ammonia

Traditionally, ammonia is manufactured using pure hydrogen produced by steam reforming in a reactor system such as that of Fig. 4.2 and a plant such as that shown in Fig. 4.4. The hydrogen is then fed together with the nitrogen to an ammonia synthesis reactor. The synthesis of ammonia:

$$N_2 + 3H_2 \rightleftharpoons 2NH_3$$

is an exothermic reaction and high yields are thermodynamically favoured by operation at low temperature and low pressures. However, since the ammonia product is required at higher pressures (ca. 30 atm.), the synthesis is carried out at this pressure in a sequence of reactors operating at a series of decreasing temperatures, the final conversion being determined by the temperature of the final reactor.[t]

Haldor Topsøe has recently introduced a new technology for ammonia production, the 'Syncor Ammonia Process', this being based on the production of hydrogen from natural gas using their 'Syncor' reactor (see

---

[t] For a full description of the operation of some typical ammonia synthesis reactors, the reader should consult the book listed under footnote 'c'.

Fig. 4.14).[u] This reactor is a variant of the more usual autothermal reforming system (Fig. 4.12). Following a pre-reforming step in which any higher hydrocarbon present in the feed gas is converted by steam reforming to syngas at lower reaction temperatures, a mixture of natural gas and oxygen (the latter produced using an air separator), together with the products of the pre-reformer and some steam, is ignited at the top of the reactor. The flame of combustion gases (now a mixture of unburnt methane, CO, CO_2 and H_2O) heats the catalyst bed at the base of the reactor up to reaction temperature. The mixture is then converted completely to syngas without the need for any additional heat input.

Following two-shift reactors, both now operated at high temperature, and CO_2 removal, nitrogen from the air separator is added to the hydrogen stream and the mixture is admitted to a three-stage ammonia synthesis reactor. Fig. 4.15 shows a schematic representation of the complete plant. A full description of all the components of the plant is given in reference of footnote 't'.

Because there is no need for external heating, the plant with the Syncor reactor (Fig. 4.16) is much smaller than the equivalent conventional plant using tubular steam reforming reactors (Fig. 4.4). The consequence is that the maximum capacity for a Syncor plant can be up to 6000 metric tons

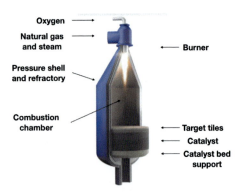

Fig. 4.14 Cutaway diagram of the Haldor Topsoe Syncor reactor. *(Diagram reproduced with permission from Haldor Topsoe's White Paper on the New SynCOR Ammonia process; www.topsoe.com. Reproduced with the kind permission of Haldor Topsoe.)*

---

[u] P.J. Dahl, C. Speth, A.E. Kroll Jensen, M. Symreng, M.K. Hoffmann, P.A. Han, S.E. Nielsen, 'New Syncor Ammonia Process', Haldor Topsoe white paper, cvr 41853816/CCM/0242.2017 (www.topsoe.com).

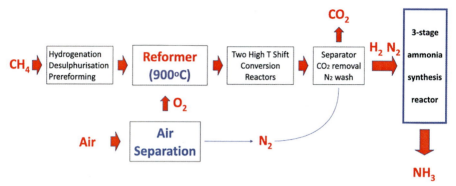

Fig. 4.15 Simplified schematic diagram of a complete ammonia synthesis plant incorporating a Topsoe Syncor reactor. *(Adapted from the white paper of footnote 'n'.)*

of ammonia per day compared with a more conventional plant for which the maximum capacity is only about 1500 metric tons per day. Another very important aspect of Syncor technology is that it operates at steam to carbon ratio of about 0.6. This is much lower than that in a conventional plant for

Fig. 4.16 An ammonia plant with a Syncor Autoformer. *(Reproduced from Haldor Topsoe white paper on Syncor Ammonia Process. (n.d.). www.topsoe.com, with kind permission. Figure kindly supplied by Haldor Topsoe.)*

which the steam/carbon ratio is about 3.0 and there is therefore a much-reduced requirement for steam handling equipment. Most importantly, as there is no requirement for reactor heating, the all-over production of CO_2 is much reduced when compared with the conventional plants involving steam reforming. Further, because of there being no need for steam recycling, the all-over site area ('footprint') required for the complete installation is also much lower. (Compare Fig. 4.4 and Fig. 4.16). Further developments related to this approach but using hydrogen produced by electrolysis are discussed in Chapter 7.

Conclusions

This chapter describes the methods used for the production of hydrogen, methanol, and ammonia. Depending on whether or not carbon dioxide capture and storage (CCS) is applied, these molecules can be termed as "Blue" or "Grey" (or even "Black" if the hydrogen is produced from coal). Use of the "Blue" products could make a very significant contribution to a reduction in total global green-house gas emissions. However, the success of such an approach will depend strongly on the success achieved in developing reliable CCS facilities. A more sustainable approach will be to make use of hydrogen and other products formed making use of renewable resources. In this context, Chapter 5 examines the use of biomass as a source of hydrogen and other chemicals while Chapter 7 includes a discussion of the developing opportunities related to the use of green hydrogen formed by electrolytic methods for transportation and chemical processing.

CHAPTER 5

Biomass as a source of energy and chemicals

Introduction

This chapter is devoted to the increasingly important topic of biomass utilisation. Biomass is formed by processes in which growth occurs by photochemical reactions of carbon dioxide and water, these giving products in which energy is stored. All the reactions of biomass which will now be discussed involve the release of some or all of this stored energy; the most stable products from biomass conversion are carbon dioxide and water formed in combustion reactions but many other less stable products are also possible. Although the various fossil fuels such as coal and oil discussed in earlier chapters were originally derived from biomass materials, the term biomass as used in this chapter refers to material originating recently, or relatively recently, from living vegetation, sources ranging from trees and grasses to seaweed and algae. Since both animals and humans depend either directly or indirectly on one or other type of vegetable matter as a source of food, the term also covers various forms of the resultant agricultural and human waste ('organic residues'). Grasses, starch crops and sugar crops can all be converted to sugars and hence to other valuable products by straightforward hydrolysis. Oil crops can be converted to biodiesel while lignocellulosic materials, lignocellulosic residues and algae can be converted by more exacting hydrolysis processes and also by fermentation or pyrolysis.[a] Fig. 5.1 shows some of these processes in a schematic form. The main chemical elements contained in all these biomass materials are carbon, hydrogen and oxygen, these resulting from CO_2 and H_2O. However, biomass sometimes also contains significant proportions of sulphur and nitrogen as well as inorganic elements that are incorporated during the growing process. As will be discussed further in later sections of this chapter, the

---

[a] F. Cherubini, G. Jungmeier, M. Wellisch, T. Wilke, I. Skiadas, R. van Ree and E. de Jong, 'Towards a common classification approach for biorefinery systems', Biofuels, Bioproducts and Biorefining, 3 (2009) 534–546.

Sustainable Energy
https://doi.org/10.1016/B978-0-12-823375-7.00004-4

Copyright © 2022 Elsevier B.V.
All rights reserved.

103

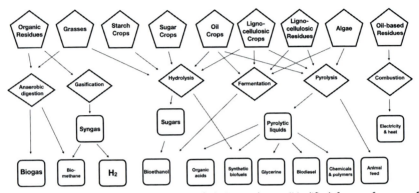

Fig. 5.1 Types of biomass and some of their products. *(Modified from reference of footnote 'a'.)*

predominant structural organic components of many biomass types are cellulose (a polymer composed of glucose units), hemicelluloses (a range of polymers composed of pentoses, especially xylose, and of other hexoses, glucose and mannose) and lignin (a heterogeneous phenolic polymer); in most cases, these are combined with lipids, proteins, starches and hydrocarbons. This chapter first discusses two long-standing uses of wood: the *combustion* of timber as a source of energy; and the production of paper from wood by pulping methods. It then covers routes by which various forms of biomass, including timber, can be converted into some important chemical products.

Wood as a source of energy and paper[b]

Biomass has been a source of energy for many centuries and the combustion of wood was the only source of energy for both heating and cooking that was used in many ancient civilisations before the use of coal became widespread. Biomass combustion still predominates for these purposes in many of the less advanced countries around the world. The combustion of wood gives rise predominantly to water and CO_2. However, there can also be significant emissions of SO_2 and NO_x and these emissions can be accompanied by significant proportions of non-combusted organic components as well as of particulates. These emissions are now known to cause severe health problems, something that is particularly problematic in developing countries where biomass combustion is used for domestic purposes, often

[b] Timber is also an important building material but that use is outside the scope of this book.

in situations where the ventilation is inadequate. Such emissions are particularly problematic in situations in which other materials such as cow dung are used as fuel. There has been a resurgence in the use of wood as a fuel, the argument being that it is a sustainable resource. The growth of biomass requires a combination of CO_2, water and sunlight and so plant growth is seen as sequestering CO_2. Hence, when the biomass is burnt, the resulting CO_2 is reckoned to be replacing that which had previously been sequestered. It is therefore commonly argued that a carbon balance is maintained as long as any material harvested for combustion is immediately replaced by establishing the growth of new plants. This argument fails to consider the whole picture and does not take into account the fact that the new growth does not immediately absorb as much CO_2 as was liberated during the combustion process. This is because the carbon accumulated in a plant or tree is stored not only in the growing vegetation but enters the surrounding soil via the roots where it accumulates in a time-dependent process as humic materials. Older growths accumulate these carbon-storing humic materials much more effectively than does new growth and so the carbon sequestering properties of new replacement plants is significantly lower than that of established plants. It will thus require many years of growth before any replacement forestation provides the same level of carbon sequestration as did the original trees. Another important factor in such a replacement programme, often ignored, is that the harvesting and planting processes invariably require energy inputs (farm transport, production and use of fertilisers, etc.). Hence, the all-over carbon balance associated with biomass combustion may be significantly negative, i.e. severely detrimental, rather than being carbon neutral. It is clear that practices such as forest clearance without replacement of the original vegetation are highly undesirable: even if the trees harvested are replaced by the growth of annual crops, such clearance decreases the allover sequestering ability of that region and contributes to significant increases in the greenhouse gas content of the atmosphere. The combustion of wood for heating purposes is therefore only justified if the wood is harvested as part of a fully sustainable forestry programme and if the predominant materials used are the result of forest thinning and pruning activities. Further, the timber that is used must be well dried before combustion or the temperature will be inadequate to provide complete combustion and hence the emissions will contain non-combusted organic molecules and particulates. The latter contaminants can be removed by fitting a catalytic converter to the stove used (see Box 5.1) but unfortunately, this technology is little used. One of the associated problems is that the catalysts of such devices need to be

BOX 5.1 Catalytic Wood-burning Stoves

Modern wood-burning stoves are constructed in such a way as to improve combustion efficiency and reduce undesirable emissions. They are generally designed for the combustion of uniform pellets with very low moisture contents that can be made from many different types of well-aged wood. The pellets can also be made from other forms of biomass such as coconut shells. A modern stove allows preheating of the air feed to the combustion chamber and good thermal emission from the stove to its surroundings. A catalytic converter can be included in the exit flue of the burner, this generally containing a Pd-containing element that enables further oxidation of unburnt components and of any CO formed in the burning process. Fig. 5.2 shows schematically the construction of such a stove. Cold air (feed, shown in blue) enters the base of the stove and is preheated as it passes the front of the stove on its way to the combustion region. The product gases leave the top of the stack of burning pellets and pass through the catalytic converter where combustion is completed utilising some of the excess oxygen of the feed. The stove in the diagram includes a baffle to the righthand side of the catalytic element that can be opened to allow the flue gas to bypass the catalyst during the start-up of the combustion process.

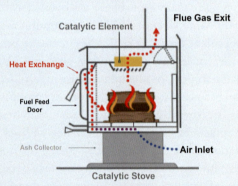

Fig. 5.2 A wood-burning stove with a catalytic converter. *(Modified from www.epa.gov/burnwise/choosing-right-wood-burning-stove#catalytic; this website gives useful information on the advantages of installing a wood stove containing a catalytic element.)*

protected from exposure to the high water vapour contents of the gases emitted in the initial combustion processes by opening the bypass baffle. The catalyst also needs to be well maintained and replaced as needed and so the effective operation of a stove of the type shown in Box 5.1 requires some care.

Biomass as a source of energy and chemicals **107**

The use of wood in paper production

The production of paper from wood has been practised for many centuries and the chemistry of the breakdown of the lingo-cellulosic structure of the wood to give the cellulose fibres ('paper pulp') for use in the paper-making process is well understood. A total of 187 million tonnes of paper pulp was produced in 2018 (see Table 5.1).[c] The paper pulp industry is a significant energy user, consuming of the order of 7×10^{18} J (7 EJ) per year (about 6% of the world's energy consumption), and it is thus the fourth largest industrial energy user worldwide.

The extraction of cellulose fibres from wood is carried out using 'pulp mills' such as those used in the Kraft process (Fig. 5.3). The raw material used in such a mill is dry wood chips, often accompanied by recycled paper (see the next section). In the initial step of the pulping process, the feed material is treated in the digester with an aqueous solution containing sodium hydroxide ($NaOH$) and sodium sulphide (Na_2S). This treatment breaks down the lignocellulosic structure of the wood chips by depolymerising the lignin species that bond together the linear fibrous cellulosic and hemicellulosic polymers of the structure, leaving them relatively intact. (See Figs. 5.4–5.6 of Box 5.2.) By-products of the pulping process include turpentine (5–10 kg per ton of pulp) and 'tall oil' soap (30–50 kg per ton of the pulp). For more information on the various pulping processes used, the reader should consult one of the references in footnote.[d]

Table 5.1 Worldwide production and consumption of paper pulp in 2018.

Region	% Production	% Consumption
Europe	25.1	27.1
North America	33.9	28.9
Asia	22.2	36.3
Latin America	16.3	5.4
Rest of World	2.5	2.3
	Total World Production 187.2 million tonnes	Total World Consumption 186.6 million tonnes

Based on Cepi statistics. (See footnote 'c').

[c] 'Key Statistics 2019' published by the European Pulp and Paper Industry (CEPI); see https://www.cepi.org/wp-content/uploads/2020/07/Final-Key-Statistics-2019.pdf. This report gives a detailed breakdown of the European paper industry.

[d] For a full description of the Kraft process, the reader should consult sources such as https://encyclopedia.pub/976; https://en.wikipedia.org/wiki/Kraft_process.

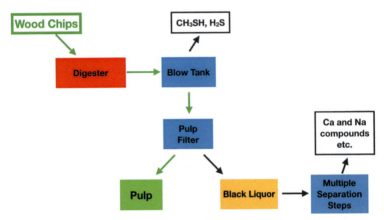

Fig. 5.3 The layout of a Kraft wood-pulping plant. The main depolymerisation reaction takes place in the digester and the pulp is then cleaned further and collected. The gaseous products are separated out and various other compounds are prepared from the black liquor, these including CaO and sodium salts. *(A schematic depiction of the plant is to be found at https://en.wikipedia.org/wiki/File:Pulp_mill_2.jpg.)*

BOX 5.2 The structure of lignocellulosic materials found in wood and other forms of biomass.

The main components of lingo-cellulosic materials are cellulose, hemicellulose and lignin. Cellulose is made up of a series of linear chains of D-glucose monomer ($C_6H_{12}O_6$) that are linked by *beta*-1,4-glycosidic bonds between successive glucose rings as shown in Fig. 5.4.

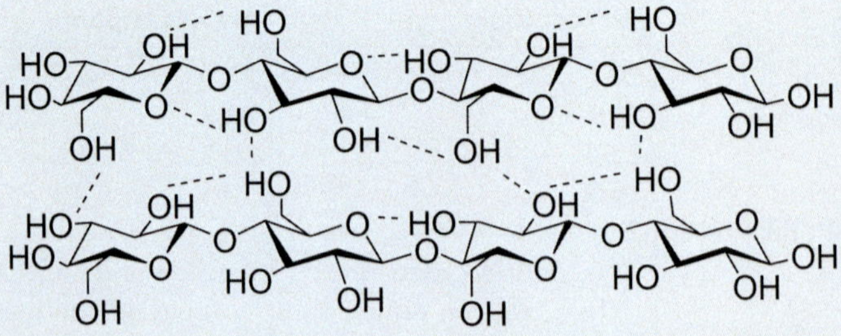

Fig. 5.4 The structure of cellulose. *(From Aqueous-phase hydrolysis of cellulose and hemicelluloses over molecular acidic catalysts: Insights into the kinetics and reaction mechanism. (2016). Applied Catalysis B: Environmental, 184, 285–298. Reproduced with the kind permission of Elsevier.)*

The dotted lines in this figure represent hydrogen bonds, the presence of which contributes to the rigidity of the linear structure; in particular, the hydrogen bond between successive glucose rings ensures the linear nature of the polymer. The degree of polymerisation depends on the source of the lignocellulose: plant cell

Continued

Biomass as a source of energy and chemicals   **109**

BOX 5.2 The structure of lignocellulosic materials found in wood and other forms of biomass—cont'd

walls generally contain cellulose chains with 5000 to 7500 glucose units while wood and cotton materials contain 10,000 to 15,000 units. Approximately 1.5×1012 tons of cellulose are produced annually.[e] As will be discussed in a subsequent section, the cellulose structure can be hydrolysed to form the individual sugars. Because of the complex structure of the lignocellulosic materials in which it is present, this hydrolysis occurs much less easily than if the chain existed alone.

Hemicellulose is also a polysaccharide but instead being composed of only the six-carbon D-glucose units as in cellulose, it is made up of a variety of different sugars including the five-carbon sugars xylose and arabinose, the six-carbon sugars glucose, mannose and galactose and the six-carbon deoxy-sugar rhamnose. The linkages between the sugar entities have some similarities to those in cellulose. Particularly important is that the structure is generally non-linear, having a number of side-groups. The composition of hemicellulose depends strongly on the plant type in which it is found; most contain predominantly xylose but soft-woods can contain predominantly mannose. Fig. 5.5 shows the structure of rabinoxylan, one of the most predominant forms of hemicellulose polymer. In this, the polymer chain is made up of the five-membered sugar, xylose, with side chains of the five-membered sugar arabinose. (The structures of some of the other varied forms are shown in the review cited for Figs. 5.4–5.6.[e]) Hemicelluloses consist of chains of between 500 and 3000 sugar units, much shorter than those in cellulose; the length of the polymer chain depends also on the degree of growth of the plant. The hemicellulose is connected to the cellulose chains by hydrogen bonds and Van der Waals forces rather than by means of chemical bonds. Hemicelluloses can also be hydrolysed to form the component sugars.

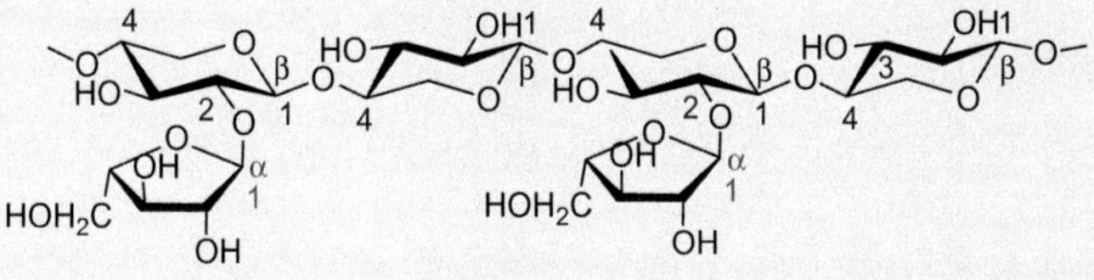

Fig. 5.5 The structure of rabinoxylan, one of the components of hemicellulose, showing the existence of side-chains. *(From Aqueous-phase hydrolysis of cellulose and hemicelluloses over molecular acidic catalysts: Insights into the kinetics and reaction mechanism. (c. 2016)., 184, 285–298. (2016). Applied Catalysis B: Environmental, 184, 285–298. Reproduced with kind permission of Elsevier.)*

Lignin, a complex organic polymer, is the third component of lignocellulose. Its composition varies from plant to plant but common to many of the structures is a phenylpropane entity. The lignin molecules provide the linkages between the cellulose and hemicellulose chains of the lignocellulose structures

Continued

110 Sustainable energy

BOX 5.2 The structure of lignocellulosic materials found in wood and other forms of biomass—cont'd

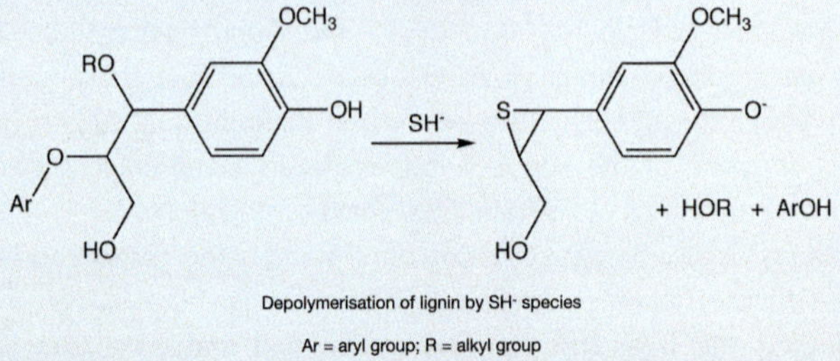

Depolymerisation of lignin by SH⁻ species

Ar = aryl group; R = alkyl group

Fig. 5.6 The net reaction for the depolymerisation of lignin using bisulphide (SH—) anions. (Ar=aryl; R=alkyl group). *(From Aqueous-phase hydrolysis of cellulose and hemicelluloses over molecular acidic catalysts: Insights into the kinetics and reaction mechanism. (2016). Applied Catalysis B: Environmental, 184, 285–298. Reproduced with the kind permission of Elsevier.)*

and are the main contributor to the strength of these structures, making them relatively stable to break down by hydrolysis under normal conditions. In the paper-making process discussed above, the preliminary hydrolysis is carried out using HS- ions, as shown in Fig. 5.6. The remaining traces of lignin are responsible for the ageing of the low-quality paper used for newspapers; the lignin must be removed completely from the pulp to be used for the production of higher-grade papers.

[e] L. Negahdar, I. Delidovich and R. Palkovits, Aqueous-phase hydrolysis of cellulose and hemicelluloses over molecular acidic catalysts: Insights into the kinetics and reaction mechanism Applied Catalysis B: Environmental 184 (2016) 285–298.

Paper recycling

Paper waste is a major constituent of municipal solid waste and although a significant proportion of this waste, about 50%, is recovered and recycled, paper waste is still the major component; see Box 5.3. It has been argued that the recycling of 1 ton of newsprint saves 1 ton of wood; the figures estimated for the reduction of energy required range from 40% to 64%. Recycling enables the cellulosic fibres recovered from the recycled paper to be extracted and reused in making new paper. However, the extraction process that is carried out in what is often a specialised pulping plant requires that any additives incorporated in the original paper-making process are extracted and that any ink residues are also removed. The recycling process often means that the cellulosic fibres are also partially disrupted by the extraction process and so the recycled fibres are often used in the preparation of lower quality products such as tissues rather than of paper

BOX 5.3 Paper recycling in the Netherlands

The recycling of paper and board products is a well-established procedure. In Europe (i.e. the EU plus Norway and Switzerland), 72% of such waste was recycled in 2019, this figure reflecting a significant increase since 1991 when only 40.3% was recycled.[3] In the Netherlands, 86% of the paper and board produced is made from recycled materials. A recent Dutch report[g] provides some very readable information on the processes involved in recycling and also discusses the structure of the Dutch industry and the types of products being produced. The majority of material recycled (85%) is termed 'coloured paper' and this is mostly used for the production of the corrugated and solid board for packaging purposes. Between 10% and 13% of the recycled material is white paper and that is re-used in the production of higher quality papers and board. The remaining 2% consists of items such as coated beverage containers that require specialised re-treatment. According to the Dutch report, approximately 15% of the paper for recycling consists of non-fibrous materials; of this, an average of 20.1% is clay, 57.6% is $CaCO_3$ and 11.2% is starch; the remaining 11.1% consists of other components such as inks, the actual proportions of these components depending on the material being processed.[7] The Dutch report emphasises the importance of careful formulation of all the paper and board that will be used commercially so that they are easily recycled. The printing processes used on the final products must also be carefully controlled, bearing fully in mind the ease with which the inks can later be removed during re-pulping; it is preferable that all the inks are water-soluble. Further, the proportion of waste that remains after re-pulping procedures have been carried out must also be recyclable or combustible and for this reason, the use of components such as PVC in paper coatings is undesirable.

[g] https://circpack.eu/fileadmin/user_upload/Recycling_of_Paper_and_Board_in_The_Netherlands_in_2019_-_final.pdf.

of the quality required for printing. Further, high-quality paper often contains additives such as china clay to provide a suitable surface texture and the clay materials separated out in such repulping activities can also be reused in other processes. As an example of such a recycling activity, the clay minerals extracted have been used for the manufacture of a cement substitute. In another example, the short cellulose fibres obtained from recycled toilet paper have been used as a component in the surface coating of bicycle lanes in the Netherlands.[f]

[f] https://www.mentalfloss.com/article/504999/netherlands-paving-its-roads-recycled-toilet-paper.

112   Sustainable energy

Non-traditional uses of biomass: First and second generation bio-refinery processes

We now consider some of the various routes outlined in Fig. 5.1 that are either currently available or are being developed for the conversion of biomass into useful products in so-called 'First and Second Generation Bio-Refineries'. Much of the biomass produced in the world is used either directly or indirectly as food for humans and animals.[h] The production of chemicals and biofuels (predominantly biogas, ethanol and biodiesel) from traditional food crops in so-called 'First Generation Biofuel Refineries' will first be discussed. The crops required for all these products are grown on arable land that might otherwise be used for food production, either indirectly for the production of animal feed or directly for human consumption. Hence, there is considerable concern regarding the competition between the production of products in first-generation bio-refineries and food production.

Organic residues and grasses

Fig. 5.7, an expansion of the left-hand section of Fig. 5.1, shows the predominant routes available for the conversion of organic residues and grasses. Organic residues are materials such as agricultural, human and food waste but they also include items such as the cellulosic fibres from paper recycling that cannot be re-used in other processes such as paper-making. In many countries, all these various residues are often either sent to landfill or are burnt for energy production in municipal incinerators. Alternative and more environmentally responsible processes are either *anaerobic digestion*[i] to form biogas or *gasification* to produce syngas and hence bio-methane or hydrogen. The biogas formed in the anaerobic digestion process contains predominantly CO, CO_2, H_2 and CH_4 and this mixture can be used directly as a source of combustion energy, e.g. in agricultural and horticultural settings where energy is required for heating purposes (animal sheds, greenhouses, etc.); using suitable anaerobic digestion conditions, the product can consist predominantly of 'bio-methane'. There is currently much activity in many countries in which anaerobic digestion is

---

[h] Human and agricultural waste are also a consequence of food production and the methods available for the treatment of such wastes are also discussed briefly below.

[i] Aerobic digestion is also widely practised as in sewerage works. However, such systems are used largely for waste disposal rather than for the generation of energy and chemicals, and any solid waste is used as a fertiliser. An exception to this practice would be the use of the solid organic residues from an aerobic sewerage disposal plant in anaerobic digestion or gasification processes as described here.

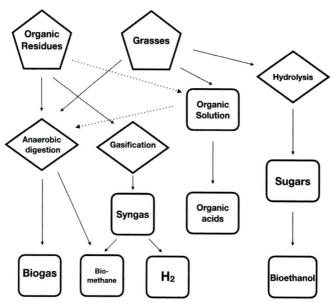

Fig. 5.7 The conversion of organic residues and grasses to valuable products. *(Based on data from the reference of footnote 'a'.)*

practised that is aimed at adding methane to existing natural gas supply networks. As the bio-methane is generally produced at or near atmospheric pressure, it can either be pressurised prior to admitting it to a high-pressure gas pipeline or it can be supplied at lower pressure directly to the consumer.

Grass is one of the most common biomass species and it is most commonly used as a feed for cattle and other ruminants, this ultimately giving rise to some of the organic residues discussed above. However, grass can also be treated to produce a solution from which organic acids can be extracted, the remaining solid residue also being used in anaerobic digestion systems. Further, grass can be hydrolysed to give a solution containing a variety of sugars and these can then be further processed to form products such as bioethanol as discussed in more detail in the following section.

Ethanol and bioethanol production

Ethanol is an important chemical that is used in many industrial applications. It is an important product of the petrochemical industry in which it

114 Sustainable energy

is formed by the hydration of the ethylene produced from oil in catalytic crackers.

$$C_nH_m \rightarrow CH_2 = CH_2 \left(+ CH_3 - CH = CH_2 \right)$$
$$CH_2 = CH_2 + H_2O \rightarrow C_2H_5OH$$

This hydration reaction requires the use of an acidic catalyst such as supported phosphoric acid at temperatures about 250°C. The majority of the ethanol formed is subsequently used in other petrochemical processes. The annual world production of ethanol by the petrochemical route, about 0.4 million tonnes per year, is approximately 4% of the total world production.

The vast proportion of the remaining ethanol now used worldwide (Table 5.2) is produced by various fermentation routes. In consequence, the product is generally described as 'bioethanol'. These methods in many cases differ little from those that have been used for centuries for beverage production. The total annual world production of ethanol by both routes is around 40 million tonnes per year, of which 77% was used as a fuel (largely for transport, see below), 8% as a component of beverages and 15% in industrial applications. As can be seen from the data in Table 5.2, the predominant producers of bioethanol are the United States and Brazil.

Bioethanol can be derived from the starch or sugars contained in many different biomaterials but it is currently mostly produced in commercial quantities from either the starch content of maize (which is the predominant feed in the US) or the sucrose content of sugar cane (the predominant feed in Brazil). Corn

Table 5.2 The major global producers of bioethanol.

Country	Production in 2019/million gallons	% of world production
United States	15,788	54
Brazil	8590	30
European Union	1370	5
China	1000	3
Canada	520	2
India	510	2
Thailand	430	1
Argentina	280	1
Rest of World	522	2
Total Production	29,000	100

Adapted from data provided by the Renewable Fuels Association (https://ethanolrfa/statistics/annual-ethanol-production/).

contains 60%–70% of starch and the remainder consists mainly of proteins (8%–12%) and water (10%–15%). (Other predominantly starch-containing plants which can also be used for bioethanol production include cereals—such as wheat, rye, barley and sorghum—or root crops such as sugar beet.) The sugar from these sources consists largely of two polymeric species: amylose and amylopectin. The former is a water-insoluble straight-chain polymer composed of alpha-glucose subunits while amylopectin is a branched-chain polymer of the same alpha-glucose subunits. Prior to fermentation of a feed such as corn, these polymeric species have to be broken down to give fermentable sugar units and this is generally carried out in an acid-catalysed hydrolysis process occurring in two steps: initial breakdown to oligomers and subsequent reaction to form the sugar monomers. (A process involving enzymatic hydrolysis is also nowadays increasingly practised.) The allover process of obtaining alcohol from a starch crop such as corn is outlined in Fig. 5.8.[j]

In the case of sugar cane and other sugar-containing crops such as sugar beet, the hydrolysis step is not necessary and so the operational costs associated with bioethanol production from these crops are significantly lower. The sucrose contained in the cane or beet is extracted by crushing them in water

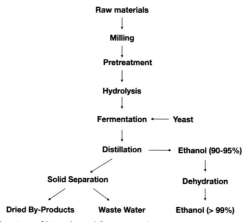

Fig. 5.8 The production of bioethanol from starch-containing plants. *(Diagram modified from 'Bioethanol: Market and Production Processes', M.J. Taherzadeh and K. Karami, 'Biofuels Refining and Performance', Ed. A. Nag, McGraw Hill, (2008) 69–123.)*

---

[j] This and many other related topics are covered in some relatively recent textbooks and collections of reviews. An example of the latter type is 'Biomass to Biofuels-Strategies for Global Industries' edited by A.A. Vertès, N. Qureshi, H.P. Blaschek and Y. Yukawa, 2010, John Wiley and Sons, Chichester.

and it is then purified prior to the enzymatic fermentation step.[k] Following the fermentation stage, after which the maximum alcohol content is about 12 wt%, the product is subjected to distillation, this resulting in an azeotropic mixture of alcohol and water containing 95.57 wt% alcohol. This product can be used as such in most applications and the solids remaining after the distillation process can be further processed, used as animal food or combusted for energy production. However, in order to obtain pure alcohol ('absolute alcohol') for use, for example, as an additive for petrol, further purification is needed. This involves either further distillation of a ternary solution containing an additive such as benzene or cyclohexane or the extraction of the water from the azeotropic mixture by the use of molecular sieve drying technology or by selective membrane separation methods.

One of the primary current uses of bioethanol, now also produced from lignocellulose (see below), is its dehydration to produce ethylene (by the reverse of the reaction shown above for the formation of ethanol from the ethylene produced by cracking crude oil). This ethylene is then used for the preparation of a variety of important chemicals and polymers as shown in Fig. 5.9. Table 5.3 shows the quantities of some of the products

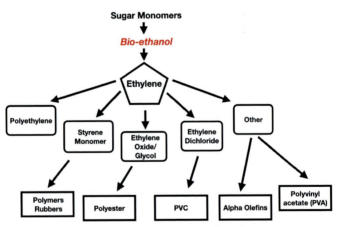

Fig. 5.9 The conversion of bioethanol to a variety of products.

---

[k] Sugars suitable for fermentation can also be produced from oil crops and from materials comprising predominantly of lignocellulose; however, the preprocessing in these cases is more complex. As we will see, the lignocellulose components of the residues from other crops such as sugar bagasse (the material remaining after sugar extraction) or various other stover materials can also be used as feedstocks for further processing.

Table 5.3 Worldwide production of some important industrial commodities from bioethanol.

Product	Saving in greenhouse gas emission/tonne CO_2 per tonne product	World capacity/ mtonne per year	Annual decrease in greenhouse emissions/m tonne per year
Acetic acid	1.2	8.3	9.6
Acrylic acid	1.5	2.9	4.4
Adipic acid	3.3	2.4	7.9
Butanol	3.9	2.5	9.6
Caprolactam	5.2	3.9	20.0
Ethanol	2.7	2.6	7.1
Ethylene	2.5	100.0	246
PHA	2.8	57.0	15.0
PLA	3.3	11.1	35.6

PHA, polyhydroxyalkanoate (an easily biodegradable polymer); *PLA*, polylactic acid (less easily biodegradable polymer formed from lactic acid).
Table constructed from data of reference in footnote 'a'.

currently produced globally from bioethanol and the savings in greenhouse gas emissions achieved by adopting a biomass-based route. The predominant commodity in this listing is ethylene, the main use of which is polyethylene manufacture.

Conversion of oil crops and oil-based residues to biodiesel and chemicals

Just as the fermentation of sugar-based crops to produce alcohol has been practised for many centuries, naturally occurring oils such as palm oil, olive oil and rapeseed oil that are easily extracted from the associated biomass have been used since early times for lighting, medicinal and cooking purposes. Oil for such uses can also be extracted from many other types of seeds and berries and there are also many uses for the residual oil-containing materials. More recently, there has been a significant interest in the use of crops for the production of energy carriers, particularly biodiesel, and also of chemicals. The crops now most commonly used for biodiesel production are rapeseed (*Brassica napus*) and soya bean (*Glycine* max).[1] The processes that will now be discussed are in direct competition

---

[1] The methods used for biodiesel production can also be applied to the treatment of used cooking oil, this making it cost-effective to collect and treat this material rather than disposing of it as waste.

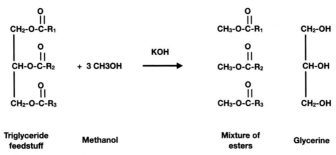

Fig. 5.10 The transesterification process.

with food production and are therefore regarded as being first-generation methods.

The oils extracted from the various sources mentioned above consist of triglycerides. As shown on the left-hand side of Fig. 5.10, these are esters of glycerine containing three different RC(O)O groupings derived from the corresponding acids. The simplest of the possible acids is acetic acid, CH_3COOH, in which case R is the CH_3 group, but in practice R is generally much larger, containing between 14 and 22 carbon atoms.[m] Hydrolysis of a glyceride in the presence of simple alcohol such as methanol results in the formation of the methyl esters of the acid groupings.[n] This 'transesterification' process, catalysed by a base such as KOH, is shown in the equation of Fig. 5.10. The mixture of esters formed in the reaction can be used as diesel fuel ('biodiesel') as its combustion properties are similar to those of conventional diesel fuel as prepared from crude oil.[o] Biodiesel can be used undiluted as fuel for conventional diesel engines but it is more frequently blended with diesel produced from crude oil sources. As is discussed later, bio-diesel is now also produced from the bio-oil formed in fast pyrolysis processes. A fuller description of the methods of production and use of biodiesel is beyond the scope of this book and the reader wishing more detail should refer to one of a wide variety of comprehensive texts currently available (see footnote 'j').

---

[m] Triglycerides also exist in animal fats; they can also be extracted from oil residues of the type referred to earlier.
[n] Ethyl esters are also produced but this occurs much less frequently.
[o] Some quantities of residual methanol in the resultant fuel can be tolerated but its concentration must not be too high since methanol addition lowers the flash-point of the diesel.

Fuels and chemicals from lignocellulosic crops

All growing matter, ranging from the wood from trees to the stems and leaves of annual crops, contains cellulose, hemicellulose and lignin (see Box 5.2) in various proportions and many of these crops are distinct from those grown for food. This section therefore discusses the non-food uses of a number of chemicals that can be generated from lignocellulosic crops in 'second generation bio-refineries'. These ligocellulosic crops include not only the wood from trees that are not used for fuel or pulping but also forest residues together with some of the residues resulting from the use of food crops such as straw and corn stover. Additionally, there is increased use of specially grown crops such as miscanthus grasses. This genus includes *Miscanthus × Gigantica*, a perennial species grown in Europe that flourishes on fertile soils but can also be grown on set-aside land unsuitable for the growth of food crops. It requires minimal use of fertiliser and can be harvested in the springtime rather than in the winter months.[P]

Some of the processes that can be used for the conversion of lignocellulosic crops and their residues (i.e. lignin plus other components of the crops) are shown schematically in Fig. 5.11 and a selection of these routes will now be discussed.

As discussed in Box 5.2, cellulose is made up of linear polymeric chains comprising exclusively of the C_6 sugar, D-glucose. In contrast, hemicellulose contains a number of different sugar varieties; the main polymeric chain of hemicellulose comprises mostly of the C_5 sugar xylose but the polymer also has side branches made up of several other different types of sugar rings. The lignin that makes up the majority of the remainder of all the fibrous parts of lignocellulosic crops acts as a 'glue', holding the cellulose and hemicellulose together, the consequence being that these structures generally resist hydrolysis ('saccharification') by dilute acids or bases. It is well established that different hydrolysis procedures can result in various different degrees of depolymerisation of each of these three components. In consequence, many different hydrolysis methods are used commercially to depolymerise lingocellulose materials, the details dependent on the desired product. As an example, the depolymerisation of the lignin component of wood used for paper-making is achieved, as already discussed, using SH^- anions

---

[P] D.J. Hayes and M.H.B. Hayes, 'The role that lignocellulosic feedstocks and various biorefining technologies can play in meeting Ireland's biofuel targets', Biofuels, Bioproducts and Biorefining, 3 (2009) 500–520.

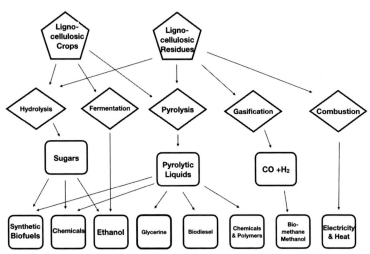

Fig. 5.11 The conversion of lignocellulosic crops and residues to useful products.

(Fig. 5.6), and the cellulose and the hemicellulose structures are relatively unaffected. Once the lignin structure is broken down, the polymer chains of the cellulose and hemicellulose components become more accessible to further hydrolysis by bases or acids and this additional hydrolysis then results in solutions containing a predominance of the constituent sugars; see Fig. 5.11. However, as the rates of hydrolysis of the chains of the polymers of cellulose and hemicellulose are significantly different, the hemicellulose structure is broken down more easily and the cyclic sugars obtained are able to break down further to form linear molecules before the hydrolysis of the cellulose is complete. Hence, many hydrolysis procedures involve two steps: the first is carried out under relatively mild conditions in order to achieve the liberation of both the lignin and hemicellulose contents; then, after separation of the sugars formed in the first step, a more severe hydrolysis process is carried out to break down the cellulose chains.[q]

Once saccharification and separation of the sugars have been carried out, the constituent saccharides can be reacted further. In particular, as they are now accessible to conventional fermentation routes, bioethanol is the predominant end product. The route for the production of bioethanol from

---

[q] When timber is the raw material being treated, the process is known as 'wood saccharification'. For more details of some of the work that has been carried out on hydrolysis procedures, the reader might consult the reference given in footnote d.

Biomass as a source of energy and chemicals **121**

lignocellulosic materials contributes a growing part of the worldwide supply as discussed in a section above (see Table 5.2), resulting in a year-by-year increase in the use of non-food feedstocks such as corn stover and sugarcane bagasse. The ethanol can then be used for the production of ethylene and hence of many other important products such as polyethylene and ethylene glycol, also discussed above (see Fig. 5.9 and Table 5.3). All these products had conventionally been made from ethylene produced by the petrochemical industry. Hence, their production from biomass has the potential to make significant reductions in the emissions of CO_2 that would otherwise have occurred.

Careful control of the hydrolysis procedure applied to lignocellulosic feedstocks can also give rise to various other useful products. One example of such a procedure is the Biofine Process[r,s] shown in Fig. 5.12 that is used to produce levulinic acid (4-oxopentanoic acid, CH_3-CO-CH_2-CH_2COOH) and formic acid (HCOOH) by reaction of the D-glucose component of the cellulose polymer and of furfural (C_4H_3OCHO) by reaction of the xylose component of the hemicellulose polymer. The levulinic acid formed can then be converted directly to a wide range of useful chemical compounds (see Fig. 5.13 and the references of footnotes 'r' and 's'). Alternatively, it can be esterified with ethanol to form ethyl levulinate, a valuable bio-diesel substitute. The formic acid, initially seen as a waste by-product, can easily be decomposed to give pure hydrogen (we will return to this subject in a later section) while the furfural produced in parallel (from the hemicellulose content of the lignocellulose) also has many uses, including its conversion to the important industrial solvent, tetrahydrofuran. The lignin fraction ends up as a char that is burnt to provide the energy needed to operate the plant.

A full-scale Biofine plant was constructed in Caserta (Italy) and this started production in about 2005 using material such as vine trimmings as feed. The plant was then acquired by GF Biochemicals (founded in 2008 and based in Geleen, The Netherlands) who later acquired the American company Segetis, a firm specialising in the downstream uses

---

[r] D.J. Hayes, J.R.H. Ross, M.H.B. Hayes and S.W. Fitzpatrick, 'The Biofine Process: Production of Levulinic Acid, Furfural and Formic Acid from Lignocellulosic Feedstocks', in Biorefineries Industrial Processes and products, B. Kamm, P.R. Gruber and M. Kamm (Eds.), Wiley-VCH, 2006.

[s] J.J. Bozell, L. Moens, D.C. Elliott, Y. Yang, G.G. Neuenscwander, S.W. Fitzpatrick, R.J. Bilski and J. L. Jarnefeld, 'Production of Levulinic Acid and Use as a Platform Chemical for Derived Products', Resources, Conservation and Recycling, 28 (2000) 227–239.

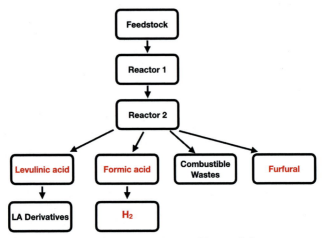

Fig. 5.12 The biofine process for the conversion of lignocellulose to various products. *(Adapted from the references of footnote 'r'.)*

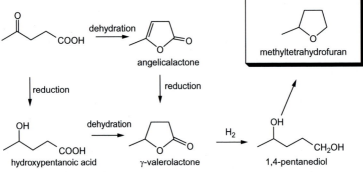

Fig. 5.13 The catalytic conversion of levulinic acid to various products. *(From 'The Biofine Process: Production of Levulinic Acid, Furfural and Formic Acid from Lignocellulosic Feedstocks', in Biorefineries Industrial Processes and products, B. Kamm, P.R. Gruber and M. Kamm (Eds), Wiley-VCH, 2006.)*

of levulinic acid. In 2020, GF Biochemicals announced a new joint venture with Towell Engineering Group of Oman for the production and marketing of various of their products, including ethyl and butyl levulinates and several ketal esters,[t] all of these intended for use as solvents.

---

[t] A ketal is derived from a ketone by replacement of the C=O group by two OH groups.

Gasification and pyrolysis of biomass
Gasification

Virtually all types of biomass can be gasified by a process similar to that used for many years for the gasification of coal. The gasification process, producing a mixture of CO, H_2 and CO_2, is carried out at a temperature of at least 700°C in the presence of oxygen and/or steam. Air can also be used as the source of oxygen but the product gas mixture will then also contain nitrogen. The syngas produced without any further purification contains additional condensable organic molecules plus methane and this makes it unsuitable for further direct processing. The removal of these contaminant species using catalytic methods has been reviewed elsewhere.[u] Syngas can also be produced directly from biomass by catalytic steam reforming, a topic also reviewed elsewhere.[v] The syngas produced in these gasification processes can be used to produce methane or methanol or used in the Fischer Tropsch process to produce motor fuels. However, the most common use of the unpurified syngas is in direct combustion for the production of electricity. The advantage of using this syngas rather than using direct combustion of the raw biomass material is that the all-over efficiency of the combustion process is greater at the higher temperature produced by syngas combustion. One of the most common uses of biomass gasification is in the treatment of biomass residues that would be otherwise difficult to process due to the inhomogeneity of these residues.

Pyrolysis

So-called 'fast pyrolysis' is now commonly used to convert both specially grown biomass and also the many forms of residue available to give a range of different products; see Fig. 5.14. The pyrolysis process involves heating the biomass or residue to a relatively high temperature (about 500°C) in the absence of and air and collecting the so-called 'pyrolysis liquid' formed. The pyrolysis liquid can then be treated chemically to give a very large range of chemical products. One of the great advantages associated with such processing is that the pyrolysis stage can be carried out on sites close to the source of biomass and the bio-oil product can then be transferred to a central processing facility for further upgrading, this being much more economical than transporting larger amounts of low-density biomass.

[u] D. Sutton, B. Kelleher and J.R.H. Ross, Review of Literature on Catalysts for Biomass gasification', Fuel Processing Technology, 73 (2001) 155–173. D.A. Bulushev and J.R.H. Ross, 'Catalysis for Conversion to Fuels via Pyrolysis and Gasification: A Review', Catalysis Today, 271 (2011) 1–13.

[v] D.A. Bulushev and J.R.H. Ross, 'Catalysis for Conversion to Fuels via Pyrolysis and Gasification: A Review', Catalysis Today, 271 (2011) 1–13.

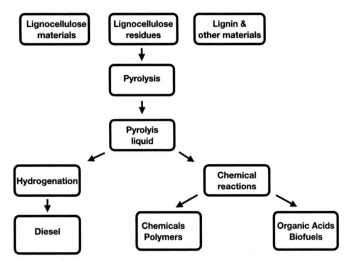

Fig. 5.14 The pyrolysis process.

One of the earliest versions of the pyrolysis process was the production of charcoal from wood, a process that has been used for many centuries. In the charcoal-burning process, the wood is heated slowly in the absence of air to produce what is essentially pure carbon, the remaining components of the wood being liberated as smoke from the kiln. However, in a modern pyrolysis system, the process is carried out using a fluidised bed reactor (see Fig. 5.15) in which the residence time is very short and from which the organic vapours emitted are collected as a 'pyrolysis liquid'. The pyrolysis process can be either non-catalytic, in which case the fluidisation is carried out using fine quartz sand as a heat transfer medium, or catalytically, in which case a suitable catalyst is included in the system (see below). The reactors used are similar to the fluid catalytic converter (FCC) units used by the petrochemical industry to convert crude oil to lower molecular weight products. Char is formed as a significant product during the pyrolysis process and is removed from the reactor together with the sand/catalyst flow; it is then combusted to provide the energy necessary for the operation of the pyrolysis process and to recirculate the sand/catalyst (Fig. 5.15).

The composition of the pyrolysis liquid obtained in the pyrolysis process depends on a very large range of factors. Of particular importance is the composition and particle size of the material being pyrolysed. However, secondary factors include the temperature of the reactor column and the residence time of the components in the reaction zone. Early versions of the fast pyrolysis process did not include a catalyst but there has relatively recently been an upsurge in the use of catalytic systems. The presence of a catalyst,

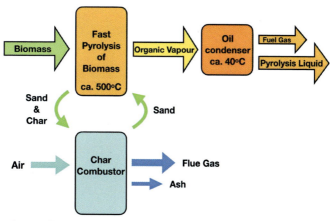

Fig. 5.15 The combustor setup of a pyrolysis unit.

either included with the sand fed to the riser system or contained in a separate catalyst bed (placed after the riser unit but before the condenser), has a very major influence on the product composition. Catalysts used in the riser reactor typically include zeolites such as ZSM-5, Y-zeolite and beta-zeolite. ZSM-5 encourages the formation of aromatic hydrocarbons during the processing of biomass materials from a wide variety of sources. Catalytic treatment after the riser reactor frequently involves hydrodeoxygenation in order to increase the H/C ratio of the products. Fig. 5.16 illustrates some of the changes that can occur in such processing.[w]

Pyrolysis oil is also a source of biodiesel by transesterification as described in a previous section and the production of biodiesel by this route is now very significant. A detailed description of all the processes that are applied to the upgrading of pyrolysis oil is beyond the scope of this book and the interested reader should consult one of a number of useful reviews of biomass pyrolysis.[x]

---

[w] A.O. Oyedun, M. Patel, M. Kumar and A. Kumar, The Upgrading of Bio-Oil via Hydrodeoxygenation, in Chemical Catalysts for Biomass Upgrading, M. Crocker and E. Santillan-Jiminez, eds., Wiley-VCH (2020) 35–60.

[x] Examples of the many reviews available on biomass pyrolysis reactions include the following: A.V. Bridgewater, Biomass Fast Pyrolysis, Thermal Science, 8 (2004) 21–49. A.V. Bridgewater, Review of Fast Pyrolysis of Biomass and Product Upgrading, Biomass and Bioenergy, 38 (2012) 68–94. N. Dahmen, E. Henrich, A. Kruse and K. Raffelt, Biomass Liquefaction and Gasification, in Biomass to Biofuels: Strategies for Global Industries, A. Vertès, N. Qureshi, H. Blaschek and H. Yukawa, ads., John Wiley and Sons (2010) 91–122. C.A. Mullen, Upgrading of Biomass via Catalytic Fast Pyrolysis (CFP), in Chemical Catalysts for Biomass Upgrading, M. Crocker and E. Santillan-Jiminez, eds., Wiley-VCH (2020) 1–33. A number of other reviews in the last two listed books also contain relevant information.

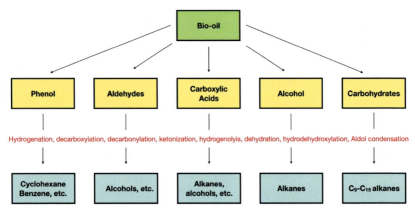

Fig. 5.16 Hydrodeoxygenation pathways for some of the compounds found in pyrolysis oil. *(Modified from the review by Oyedun et al. (Footnote w).)*

Other sources of biomass
Seaweed and algae

Seaweed and algae have long been used as a source of food, chemicals and energy. There are many different forms of both, ranging from kelps to micro-algae. The world production of seaweed exceeds 2 million metric tons a year, with the predominant usage being in China and France. The total global annual production of all aquatic plants exceeded 30 million tonnes in 2016.

Seaweed (macro-algae) has been used for centuries as a fertiliser and soil-enhancing component. For example, in rocky areas of the West of Ireland it was used in combination with sea-shore sand to form an artificial soil in which potatoes could be grown; it is now sold in powdered form as a soil enhancer to promote plant growth. Seaweed 'burning' under oxygen lean conditions similar to those used for charcoal production was a means of producing a cement for construction purposes; straightforward combustion was also used for heating and cooking purposes. Seaweed is used in many countries as a food and a series of hydrocolloids used in food products such as alginates, agar and carrageen can be produced from it. Alginates are also used in medicinal products. The growth of seaweed, in common with that of all plants, consumes CO_2 and also produces oxygen, thus contributing significantly to the reduction of greenhouse emissions. During its growth, seaweed also consumes nutrients such as ammonia, nitrate and phosphate anions as well as iron and copper cations and as a result it provides a filtering action on reefs. In China, seaweed is used to purify phosphate effluents.

It has been proposed that 'sea afforestation' could be used as a method of removing carbon emissions. The harvested material would then be treated in anaerobic digesters, producing biogas containing 60% CH_4 and 40% CO_2. The CH_4 can be used as a fuel and the CO_2 collected and stored. It is claimed that 9% afforestation of the world's oceans would remove more than the annual global emissions of CO_2. Such an approach would clearly require efficient methods to cultivate and then harvest the seaweed and this would be a serious barrier to its adoption.[y]

Microalgae

Microalgae are microscopic species with sizes ranging from several micrometres to several hundred micrometres. Unlike normal plants, they do not have roots, stems or leaves. A significant proportion of the world's microalgae are found in the oceans, rivers, lakes and ponds; a well-known example is the buildup of algal blooms in relatively static areas of water. These algae are particularly effective in photosynthesis and are responsible for the consumption of a very large proportion of the CO_2 used in global photosynthesis processes, producing much of our atmospheric oxygen. Microalgae contain a very wide range of potentially important components and are already used for the production of low-volume high-value products such as pharmaceuticals and cosmetics. There is an increased interest in cultivating them as a source of oils for use in the production of bio-fuels. Cultivation can be carried out in either marine or freshwater systems. As with seaweed, one of the main problems is associated with harvesting and the main sources are in specially-constructed ponds in which the movement of the water is carefully controlled. They can also be grown in the equivalent of greenhouse conditions (see Fig. 5.17) in which the temperature, the amount of light and the water circulation can be carefully controlled.

Chitin

The final source of biomass to be considered is *chitin*, one of the main components of the cell walls of fungi and is also found in sea shells, fish scales and

---

[y] Many web entries are available to describe the use of seaweed, e.g. https://en.m.wikipedia.org/wiki/Seaweed; https://seaweed.ie/uses_general/.

128 Sustainable energy

Fig. 5.17 A photo-bioreactor used for the growth of micro-algae. *(From https://commons.wikimedia.org/wiki/File:Photobioreactor_PBR_4000_G_IGV_Biotech.jpg)*

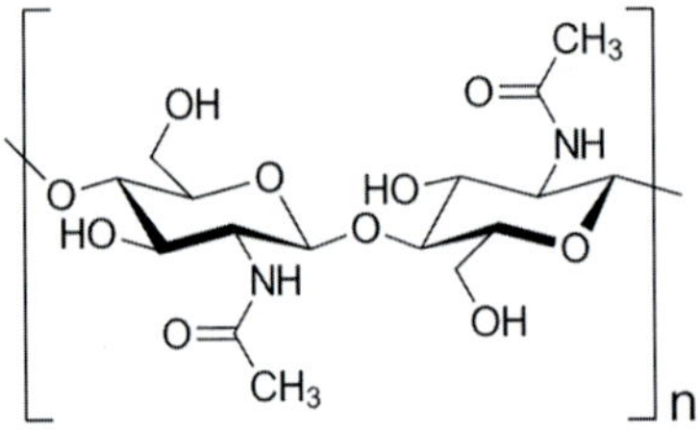

Fig. 5.18 The polymeric chain of chitin. *(From https://commons.wikimedia.org/wiki/File:Chitin.svg.)*

insect shells. It is a long-chain polysaccharide with a structure similar to that of cellulose but with units made up of *N*-acetylglucosamine. This polymer forms microfibrils that provide strength to the shells in which it is present (Fig. 5.18). It has a range of applications, from acting as a fertiliser and plant conditioner to being used in paper-sizing applications and food processing.

This chapter would be incomplete without a discussion of the organic matter contained within soil: 'soil organic matter'. Its importance in relation to the growth of all types of biomass is discussed in Box 5.4.

BOX 5.4 Soil organic matter

Soil organic matter (SOM) contains organic species such as the microorganisms, bacteria and fungi, these being stored in sub-surface soil. Although SOM does not provide a source of energy, our crops would not grow without its presence. The global quantity of this organic material exceeds by a factor of three the total amount of organic material above the surface of the soil. SOM is the 'cement' that binds the inorganic content of the soil (clays and various hydroxides, especially those of iron and aluminium) and this combination of colloidal organic and inorganic particles is the origin of soil fertility. It is important to recognise that when soils are in long-term (especially monoculture) cultivation, SOM degradation occurs. The most fertile soils of the world are being so over cultivated and has been estimated that after a further 50 to 100 crop cycles from now, the SOM of these soils will be so depleted that the soil structure will be degraded and its fertility lost. It is therefore essential that attention is also given over the next century to ensuring that SOM levels are replenished so that it will remain possible to supply food for the expanding world population.

The composition of SOM is quite complex and research on the subject is quite extensive. Extraction and fractionation of the components have been a major challenge. Techniques developed recently involve exhaustive extractions with the aqueous base at increasing pH values, followed by further extraction using an aqueous base containing urea (6M) and finally extraction of the dried residues with a solution of dimethyl sulfoxide (DMSO) and concentrated sulphuric acid (6%). This sequence allows the isolation of up to 95% of the components of SOM, each with different polarities. The more polar components isolated in the aqueous systems are enriched in organic acid functionalities and include polysaccharides, peptides and humic substances (these predominating). The generally dominant low polarity humin materials, composed of long-chain fatty acids, waxes, cuticular materials, cutin, cutan, suberin and suberan are isolated in the DMSO-acid medium.

Concluding remarks

It is clear that while biomass has been used for many centuries for a wide range of applications, there are many novel applications that are still in development which have the potential to contribute very significantly to reductions in the all-over emissions of greenhouse gases. In so doing, they can contribute to greatly improved sustainability. Many of the applications currently available are relatively little used. Hence, in common with many of the other topics discussed in this book, they require much more attention, needing substantial investment from both governments and industrial operators. As just one example, the oil industry has the potential to diversify much more extensively into the provision of biofuels. However, that would require that both political support and legislative pressure are given. Many of the areas discussed in this chapter are also in need of increased levels of basic research.

CHAPTER 6

Transport

Introduction

Transport is one of the main contributors to global greenhouse gas emissions (see Fig. 1.6). Much attention is therefore currently being given to methods by which these emissions can be reduced. This chapter starts by tracing some of the history of the development of various forms of transport currently used, almost all of which being based on the use of internal combustion engines.[a] Not only is road transport for both personal and commercial applications considered but the use of air and sea transport is also discussed. The chapter then discusses each type of transport in more detail, considering methods of improving operational efficiencies that are currently being introduced or have been proposed for future application. As it is likely that conventional systems will continue to be used for some years to come in at least in some applications, the important topic of the control of emissions from conventional automobiles is also discussed.

Historical development of mechanically driven transport

Electrical vehicles. Our ancestors relied for many centuries on horse-driven transport. However, in the late nineteenth century, following the development of the steam engine and the widespread establishment of the railways, the first efforts to produce motorised vehicles took place. Some efforts were made to create steam-driven transport. However, the first successful vehicles were electrically driven. Electricity had been known to exist for many centuries but it was the experiments by Franklin around 1752 that it began to be understood. Michael Faraday then showed in 1831 how electricity could be generated and this was followed by the introduction of the light bulb and the gradual electrification of our cities and towns. It was therefore to electricity that the early innovators in transportation turned.

---

[a] An internal combustion engine depends on the combustion of a fuel within the engine. This is distinct from the external combustion that occurs in steam engines used for railway traction and also for the operation of machinery (Chapter 1).

Sustainable Energy
https://doi.org/10.1016/B978-0-12-823375-7.00009-3

Copyright © 2022 Elsevier B.V.
All rights reserved. **131**

Fig. 6.1 The Groß-Lichterfeld tram (1882). *(https://upload.wikimedia.org/wikipedia/commons/9/90/First_electric_tram-_Siemens_1881_in_Lichterfelde.jpg.)*

The first operational electrical vehicles were trams. The first public tramway was opened in Lichterfeld near Berlin (Germany) in 1881, thus preceding the introduction of the first automobile in 1885 (see below). Fig. 6.1 shows a photograph of a tram from this system taken in 1882. It was constructed by Siemens and was 5 m long, weighing 4.8 t; it travelled at up to 40 km h^{-1} and carried 20 people at a time. This tramline initially operated using electrical current supplied through the rails but the system was modified in 1883 to use overhead wires. Similar tram systems were installed in many countries and there still exist a number of such tramways, one of the oldest of these being the Volk's Electric Railway in Brighton (UK) that was first opened in 1883. Tram systems are limited by having to have permanent track installed and such tracks can cause significant problems, as encountered for example by the large cycling population in Amsterdam (The Netherlands), a city that is still serviced by a significant and very efficient tram network; this system first operated in 1875 using horse-drawn trams but it was converted to electrical operation between 1900 and 1906.

The problem of having to have rails was circumvented in many cities by introducing trolley buses drawing the required current through moveable poles as illustrated in Fig. 6.2. The first trolley buses were introduced in Berlin in 1882 and this was soon followed by services in other German cities. The first trolley-bus service in the UK was opened in Bradford in 1911 where it operated until 1972. Belfast (Northern Ireland) opened its system in 1936, this operating until 1968; a typical trolley bus from the Belfast operation is shown in Fig. 6.3[b] Trolley buses are still in operation in many cities

---

[b] Many of the road surfaces in Belfast were very smooth and the trolley buses were very silent; one of the reasons given for their disbandment was based on safety grounds as pedestrians often did not hear their approach.

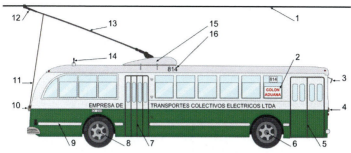

Fig. 6.2 The electrical connection of a trolley bus. 1: parallel overhead wires; 2: destination sign; 3: rear-view mirror; 4: headlights; 5: entry doors; 6: direction (turning) wheels; 7: exit doors; 8: traction wheels; 9: decoration; 10: trolley retractors; 11: pole rope; 12: contact shoes; 13: trolley poles (current collector); 14: pole storage hooks; 15: trolley pole housing; 16: bus number. *(https://en.wikipedia.org/wiki/Trolleybus.)*

Fig. 6.3 The last Belfast trolley bus to operate in 1968. *(https://en.wikipedia.org/wiki/Trolleybuses_in_Belfast.)*

throughout the world, particularly in the USSR, and there is currently a movement to reintroduce them in others.

To avoid the need for power cables, battery-operated vehicles were also developed but these had only limited success in the early attempts. Early electric buses operated using rechargeable batteries stored under the front seats and had limited ranges. Fig. 6.4 shows an electric bus developed by Edison that dates from 1915; this operated with Ni–Fe batteries. Other types of electric vehicle developed included the battery-driven milk float used for the delivery of milk and bread. Milk floats, which first appeared in use in the UK in 1889, could travel between 60 and 80 miles on one charge. Other uses of electric vehicles included road-sweepers and dust carts. We will return to

Fig. 6.4 Edison electric bus from 1915. *(https://upload.wikimedia.org/wikipedia/commons/6/6e/Edison_electric_bus_from_1915.jpg.)*

the topic of electrical traction below when discussing modern electrically operated vehicles.

Vehicles with internal combustion engines. Although there had previously been several attempts to build engines relying on combustion, for example using mixtures of hydrogen and oxygen (Francois Rivaz, 1807) or burning towns gas (Samuel Brown, 1824), the first successful spark-ignition internal combustion engine, fuelled by coal gas, was invented and patented by Jean Lenoir in 1858.[c] However, it was not until 1876 that Nikolaus Otto invented and later patented a four-stroke engine operating with gasoline that he incorporated in a motor cycle. This design became the basis of all four-stroke engines. In 1885, Gottlieb Daimler built a two-wheeled vehicle ('reitwagen') with a four-stroke engine comprising of a vertical cylinder and a carburettor, the prototype for the modern engine; the following year, he adapted a stage-coach to be the first four-wheeled vehicle, this having the same type of internal combustion engine. That same year (1886), Karl Benz received a patent for the first gasoline-fuelled vehicle, a tricycle (see Fig. 6.5); he went on in 1889 to establish Benz et Cie, this company becoming the world's largest company producing automobiles by 1900. In the same period, Daimler founded the company Daimler Motoren-Gesellschaft to build and market his designs and in 1901 this company launched the Mercedes, this having been designed by Daimler's partner, Wilhelm Maybach.

The production of automobiles in the United States was dominated in the early years by Ford and an early Ford car is shown in Fig. 6.6. First

---

[c] http://www.energybc.ca/cache/oil2/inventors.about.com/library/weekly/aacarsgasa6fc5.htm.

Fig. 6.5 The original Benz Patent Motorwagen (1885). *(https://en.wikipedia.org/wiki/Benz_Patent-Motorwagen.)*

Fig. 6.6 A Ford Model-T in Geelong, Australia for the launch in 1915. *(http://www.slv.vic.gov.au/pictures/0/0/0/doc/pi000357.shtml.)*

launched as early as 1908, it was the most widely used four-seater car of its era and is considered to have been one of the most influential designs of the twentieth century. Automobile production in the US tended towards large vehicles with very high petrol consumptions, this being related to the low cost and easy availability of gasoline. A typical American car from mid-twentieth century is shown in Fig. 6.7[d]

In Europe, there were equivalent developments of large comfortable cars, examples being those produced by Mercedes Benz and Daimler mentioned above. However, the general preference was for smaller vehicles. The engines of these cars typically had higher compression ratios and this gave greater manoeuvrability on what were generally more congested roads. Following the Suez crisis and the resultant petrol shortages, the preference for

---

[d] The author was once a part-owner of a Buick of this vintage. It was roomy and comfortable but relatively unmanoeuverable.

Fig. 6.7 A Buick Super from 1957. *(https://upload.wikimedia.org/wikipedia/commons/5/50/Buick_Super_1957.jpg.)*

Fig. 6.8 The first Morris Mini-Minor, now housed in the British Motor Museum. *(https://upload.wikimedia.org/wikipedia/commons/2/2f/Morris_Mini-Minor_1959_%28621_AOK%29.jpg.)*

smaller more compact vehicles became more emphasised. This resulted in the development of even smaller and more efficient vehicles such as the Mini series, first introduced in the UK in 1959 by Morris and Austin; see Fig. 6.8.

Vehicles with diesel engines. The first diesel engines were used in stationary applications and were only later used in vehicle propulsion. Rudolf Diesel first patented the concept of the diesel engine in a US patent application lodged in 1895 (granted 1898, No. 608845), having previously lodged another application in 1892 based on an incorrect description of the principal of operation.[e] Fig. 6.9 shows his successful prototype that ran for a total of 88 revolutions in February 1894. Diesel worked with both

---

[e] There was considerable controversy at the time as to whether or not Diesel's ideas were novel but it is now generally accepted that he was the originator of the modern diesel engine.

Fig. 6.9 Experimental diesel engine from 1894. MAN-Museum, Augsburg. *(https:// upload.wikimedia.org/wikipedia/commons/c/c7/Experimental_Diesel_Engine.jpg.)*

Krupp in Essen and Machinenfabrik Augsberg and this collaboration led to the development of diesel-powered machines that were used in many larger applications in manufacturing facilities.

The first fully functional diesel engine, completed in 1896, is shown in Fig. 6.10. This engine was rated at 113.1 kW and had an efficiency of 26%. Subsequent developments that occurred quite rapidly included the construction of

Fig. 6.10 Historical diesel engine in Deutsches Museum. *(https://commons.wikimedia. org/wiki/File:Historical_Diesel_engine_in_Deutsches_Museum.jpg.)*

Fig. 6.11 The Mercedes-Benz 260D released in 1936. The earliest diesel was launched in 1933 by Citroen. *(https://en.wikipedia.org/wiki/History_of_the_diesel_car.)*

ocean-going ships and submarines powered by diesel engines and this was followed by the construction of tractors and lorries.

The first car with a diesel engine appeared in 1929 and various companies in Europe and in the United States started to produce different models in the 1930s. As a typical example, Fig. 6.11 shows a Mercedes Benz from 1936. Improvements in the design of the diesel engine led to the high-speed engines launched by Perkins and Chapman in 1932[f] and these were soon being used in racing cars and other applications such as tractors and farm machinery. Four diesel engines were used to power the Hindenburg airship. As will be discussed further below, a diesel engine operates with greater efficiency than does an Otto engine. It can also work efficiently with a variety of fuels, these including biofuels of the type described in Chapter 5. Because the ignition of the fuel occurs at relatively low temperature, the risk of catching fire is much lower than for an engine operating with spark ignition. As they have no ignition systems, the reliability of diesel engines is also markedly greater than of Otto engines.

The efficiency of internal combustion engines. Both Otto and diesel engines are classed as being internal combustion engines since the fuel is ignited within the cylinders of both types of engine and work is done by the expanded gaseous combustion products. (This contrasts with the operation of the *external* combustion engines of the type discussed in Chapter 1.) The operation of the Otto engine is now described and this is followed by a brief discussion of the operation of the diesel engine.

---

[f] See: https://en.wikipedia.org/wiki/Perkins_Engines.

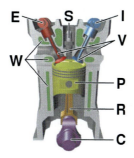

Fig. 6.12 The structure of an internal combustion engine. (C—crankshaft; E—exhaust crankshaft; I—inlet crankshaft; P—piston; R—connecting rod; S—spark plug; V—valves; W—water-cooling jacket. https://en.wikipedia.org/wiki/Internal_combustion_engine.)

The Otto engine. The operation of a typical four-stroke internal combustion engine is illustrated in Fig. 6.12, the form having changed little in the period since the original development of the Otto engine.

Each cylinder of the engine, which is water-cooled, contains a piston (P) attached to the crankshaft (C) of the engine by a connecting rod (R). As the piston first descends, it draws fuel (normally gasoline, although it could also be a gaseous molecule such as hydrogen or methane) into the cylinder through an inlet valve operated by the inlet crankshaft (I). When the fuel is gasoline, typically a non-aromatic hydrocarbon (C_nH_{2n+2}), it is vaporised due to the high temperature achieved in the cylinder and remains in vapour form while it is compressed by the piston while rising to the top of the cylinder once more. At that moment, the spark plug is fired, this causing the combustion of the gasoline vapour and resulting in a significant increase in the pressure in the cylinder due to the formation of water and CO_2:

$$C_nH_{2n+2} + O_2 \rightarrow nCO_2 + (n+1)H_2O$$

as well as from the temperature increase due to the evolution of the heat of combustion. The product gas forces the piston to descend once more. Finally, the piston rises again, this time with the exhaust valve open so that the gases pass to the exhaust system. The timing of the inlet and outlet valves as well as of the ignition spark is critical to efficient operation.

The efficiency of an Otto engine cannot exceed that predicted by the theoretically ideal Carnot cycle discussed in more detail in Box 6.1. The efficiency of such a cycle is given by:

$$\eta = W/Q_2 = (Q_2 - Q_1)/Q_2 = 1 - T_1/T_2$$

The energy to drive the motor arises from the work done by the expansion of the combustion products during the second cycle described above. The maximum efficiency achieved during this cycle is dependent on the upper and lower temperatures of the cycle. In practice, there are substantial losses of energy to the surroundings and the maximum theoretical efficiency is never achieved. Hence, on average, only 40%–45% of the energy supplied to such an engine is converted into mechanical work and used to drive the vehicle. A large proportion of the energy lost from the system is associated with heat released to the environment through the cooling system. Various

BOX 6.1 The Carnot cycle

Fig. 6.13 shows the idealised form of the reversible Carnot cycle. This cycle determines the upper limit of the efficiency of the cylinder of a thermodynamic engine such as that of the second cycle of the Otto design described in the main text, this process depending on the conversion of heat into work.[9] It can equally well apply to an engine working in the opposite direction as in a refrigeration system.

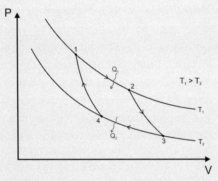

Fig. 6.13 The Carnot cycle for an Otto engine. *(https://upload.wikimedia.org/wikipedia/commons/0/06/Carnot_cycle_p-V_diagram.svg.)*

The steps in can be described as follows:

Step 1. At this stage, which corresponds to that when ignition has occurred and the product gas is compressed at the top of the cylinder at high pressure, the gas expands isothermally (at constant temperature, T_1) by moving the piston and therefore does work against the surroundings. During this stage, heat (Q_1) is being transferred to the gas from the hot surroundings. The entropy of the gas, ΔS_1, is thus increased by Q_1/T_1.

Step 2. The gas is now insulated against loss or gain of energy to or from the surroundings and expands adiabatically, moving the piston further and losing

Continued

Transport **141**

BOX 6.1 The Carnot cycle—cont'd

internal energy equivalent to the additional work done on the surroundings. There is no entropy change in this step.

Step 3. The gas is now in contact with the surroundings at the lower temperature T_2. The surroundings do work on the piston, pushing it back towards the top of the cylinder. This causes the transfer of heat (Q_2) out of the cylinder and there is a corresponding decrease of the entropy, ΔS_2, given by Q_2/T_2.

Step 4. The gas is again insulated from loss or gain of energy and the surroundings do work on the piston, pushing it back to the starting point of the cycle. Again, there is no entropy change associated with this step.

Hence, when the cycle is complete,

$$\Delta S_1 = \Delta S_2 \text{ or } Q_1/T_1 = Q_2/T_2$$

The efficiency, η, of the cycle is given by:

$$\eta = \text{work done}/Q_H$$
$$= (Q_H - Q_C)/Q_H$$
$$= 1 - T_1/T_2$$

The efficiency therefore depends on the temperature difference achieved: the higher that T_2 is compared with T_1, the higher is the efficiency. In practice, complete isolation of the cylinder containing the piston from its surroundings can never be achieved and there will also inevitably be losses of energy due to frictional forces.

[g] For a full description of the Carnot cycle and it applications, the reader should consult web sources such as https://en.wikipedia.org/wiki/Carnot_cycle; https://en.wikipedia.org/wiki/Perkins_Engines.

ways to recover some of the waste heat have been developed, such as the use of a supercharger to introduce compressed air into the cylinder, thereby increasing the efficiency of the process. Efficiency is also improved if the compression ratio is relatively high, thus causing increased engine temperatures, but that requires higher octane fuel to avoid pre-ignition or engine 'knocking'. For this reason, the high compression ratio cars used in Europe and running on relatively high octane fuel generally have better petrol consumption behaviour than the larger vehicles with lower compression ratios and more sluggish performance that have been favoured until relatively recently in the United States.

The diesel engine. The diesel engine operates on a completely different principle to that of a four-stroke Otto engine. Ignition in the cylinder of a diesel engine is brought about by using highly compressed hot air rather than a spark and uses a very different fuel. The cycle which occurs is shown in Fig. 6.14.

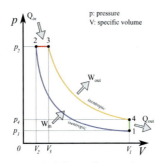

Fig. 6.14 The Diesel Cycle. The cycle follows the numbers 1–4 in a clockwise direction. The horizontal axis is volume of the cylinder. In the diesel cycle, the combustion occurs at almost constant pressure. In this diagram, the work that is generated for each cycle corresponds to the area within the loop. *(https://en.wikipedia.org/wiki/Diesel_engine#/media/File:DieselCycle_PV.svg.)*

The cycle begins at *point 1* where air is introduced into the cylinder of volume V_1. It is then compressed to volume V_2 when the pressure is now p_2, this compression step heating the air and providing energy Q_{in}. At that stage (*point 2*), the diesel fuel is injected into a space at the top of the cylinder in such a way that the liquid is broken down into small droplets and distributed evenly throughout the void space. Molecules of the fuel evaporated from the droplets then ignite spontaneously, the rate at which this occurs depending on the taste of evaporation, until all the fuel is combusted at *point 3*. (The combustion step occurs suddenly and causes the knocking sound that is characteristic of the diesel engine.) When combustion is complete at what is essentially constant pressure, the product gas expands and the piston descends once more to *point 4*.[h] The exhaust valve is opened at 4 and the product exhaust gas is then expelled, there being a decrease of pressure upon returning to *point 1*. An additional cycle that is not shown then follows during which the exhaust gases are expelled from the cylinder and a fresh charge of air is introduced before the combustion cycle is repeated. The valve arrangements of the diesel engine are similar to those in the Otto engine shown in Fig. 6.12.

The high level of compression used in a diesel engine gives a very significant improvement in efficiency when compared with the Otto engine. For this reason, the fuel consumption figures for diesel engines always exceed those for spark ignition engines of equal capacities; see Box 6.2. However, although the diesel engine has a theoretical efficiency of 75%, this

---

[h] The term isentropic shown in Fig. 6.14 refers to an idealised thermodynamic process that is both adiabatic and reversible; there is no transfer of heat or matter.

BOX 6.2 Fuel consumption figures and CO_2 emissions

In Europe, the most efficient petroleum-fuelled cars are claimed to give petrol consumptions approaching 60 miles per imperial gallon while the most efficient diesel-powered cars can give significantly higher efficiencies: greater than 70 miles per imperial gallon. In contrast, the larger cars favoured in the US generally give much worse fuel consumptions; a recent report shows that the average vehicle produced by the Ford Motor Company has a consumption of 22.5 miles per US gallon (27 miles per imperial gallon). To convert a figure in miles per gallon (mpg) to actual CO_2 emissions (g/km), for a *petrol fuelled car*, divide 6760 by the mpg figure and for a *diesel car*, divide 7440 by the mpg value; for two cars each doing 40 mpg, the values are 169 g/km (petrol) and 189 g/km (diesel). A simple rule of thumb is the following: the combustion of one US gallon of gasoline (this containing about 87% carbon) causes the emission of 20 pounds of CO_2. (www.climatekids.nasa.gov).

is never achieved as a result of heat losses through the cylinder walls and elsewhere in the system. A modern diesel car in practice has an efficiency up to about 43% while larger engines such as those in trucks and buses have slightly higher efficiencies (45%). The diesel engine has a number of operational advantages over the spark–ignition engine, including the fact that they tend to be more efficient at low loads and in consequence are often used for vehicles making short journeys with frequent stops. As is shown in later sections, their emissions are also in principle more easily controlled than those from Otto engines.[i] Further, they can be operated using a variety of different types of fuel, these including biofuels (Chapter 5). For further details of the operation of cars with diesel engines, the reader should consult articles such as https://en.wikipedia.org/wiki/Diesel_engine.

Exhaust emission control

Fig. 6.15 shows a typical layout of a modern car, this being largely self-explanatory. Apart from the internal combustion engine, essential features include a fuel injection system and an electronic control module that provides full control of all aspects of the operation of the vehicle. The fuel tanks are mostly sited under the rear seats and there is therefore a fuel line running

---

[i] Although, as discussed in a previous section, European cars tend to operate at higher compression ratios that their US counterparts, the level of compression is still lower than that in the diesel engine. This is because the use of any higher compression ratios would lead to pre-ignition and uneven operation.

144 Sustainable energy

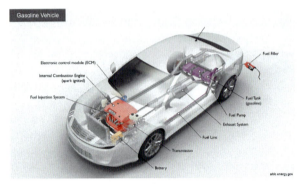

Fig. 6.15 A gasoline vehicle. *(https://afdc.energy.gov/vehicles/how-do-gasoline-cars-work.)*

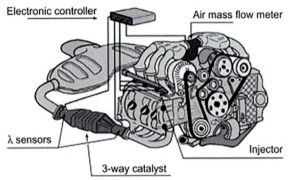

Fig. 6.16 The positioning of the catalytic filter in an internal combustion automobile engine. *(From (2003). Catal. Today, 77, 419–449, reproduced with kind permission of Elsevier.)*

under the main passenger compartment. The other very important feature of all vehicles is the exhaust system that is also found under the vehicle. This section describes in outline the methods that are used to control engine emissions.[j]

Modern vehicles with internal combustion engines always include a catalytic emission control system and this is placed between the exhaust manifolds of the engine and the silencer unit as shown in Fig. 6.16. An electronic controller is included in the system that is linked to two sensors, one before and one after the catalyst unit, and a control unit to adjust the flow of air to

---

[j] This section only gives a rudimentary description of the operation of a catalytic exhaust converter. A more detailed description of the catalysts used is to be found in Contemporary Catalysis - Fundamentals and Applications, Julian R.H. Ross, Elsevier (2019).

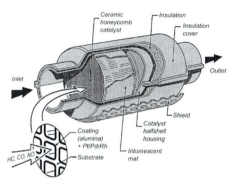

Fig. 6.17 The arrangement of the catalytic converter. *(From (2001). Appl. Catal. A, 221, 443–457, reproduced with kind permission of Elsevier.)*

the engine. The structure of a catalytic converter unit is shown schematically in Fig. 6.17. The catalyst support material, which may be in the form of either a metallic foil or ceramic honeycomb structure such as those illustrated in Fig. 6.18, is surrounded by insulation material held within a cylindrical container.

The chemical reaction occurring within an internal combustion engine of either the Otto or diesel type is the total oxidation of the hydrocarbon fuel to produce CO_2 and water. However, the product gases inevitably contain traces of unburnt hydrocarbon fuel (or lower molecular weight fragments formed from the fuel) plus carbon monoxide and the oxides of nitrogen, N_2O, NO and NO_2, the latter mixture described for convenience as NO_x. (The proportion of N_2O is generally very low.) The quantities of each of these emissions depend on the operating conditions. Fig. 6.19 shows very qualitatively how their concentrations depend on the air/fuel ratio in the

Fig. 6.18 Typical honeycomb supports used in catalytic converters; *left*: metal foil; *right*: ceramic monolith. *(From (2011). Catal. Today, 163, 33–41, reproduced with kind permission of Elsevier.)*

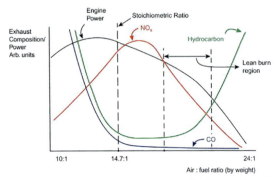

Fig. 6.19 The effect of air-fuel ratio on the operation of a four stroke Otto engine. *(From Julian R.H. Ross, Contemporary Catalysis, 2019, reproduced with kind permission of Elsevier.)*

feed to the engine; also shown in this figure is the engine power as a function of the air/fuel ratio. The stoichiometric ratio value of 14.7:1 corresponds to the optimum combustion conditions. (Below this value, the atmosphere is effectively a reducing one and above it is an oxidising one.) CO is formed at lower air/fuel ratios when there is insufficient oxygen to provide complete combustion but drops off to a relatively low level at higher ratios. The hydrocarbon component in the exhaust follows a similar path to that of CO at low air/fuel ratios but rises again at values well above the stoichiometric ratio due to there being uneven combustion of the fuel under these conditions, this being generally associated with 'knocking' of the engine. The NO_x is formed by gas-phase reactions between the nitrogen and oxygen of the feed gas at the high temperatures in the ignition stage of the cycle, the reaction being most pronounced when the temperature is at its highest near the stoichiometric composition. The engine power produced is also shown, this being at its maximum slightly below the stoichiometric air/fuel ratio; the power drops off significantly at higher values, particularly in the so-called 'lean-burn' region.

Strict air-purity control regulations were introduced in many countries in the second half of the last century. Prior to the introduction of these environmental controls, many car manufacturers, particularly in the US, produced engines that operated in the lean-burn region of Fig. 6.19 with which the NO_x emissions are relatively low and the formation of CO is negligible. However, with the introduction of the new strict regulations, the car manufacturers moved towards the production of engines operating with an air/fuel ratio slightly below the stoichiometric ratio so that the exhaust gas contains a mixture of the CO, hydrocarbon and NO_x. (This value was still above that corresponding to maximum engine power in Fig. 6.19 where the emission of CO and unburnt hydrocarbon is very high.) So-called three-way catalysts, generally containing the precious metals Pt, Pd and Rh as discussed above, were developed to oxidise the unburnt hydrocarbon and CO while at the same time reducing the NO_x component to nitrogen gas. For further details, see for example the reference of footnote 'j'.

With increasingly exacting environmental requirements, a number of car manufacturers returned to the concept of using lean-burn engines, these operating under oxidising conditions corresponding to the region with the higher air–fuel ratios shown in Fig. 6.19. In this region, both the CO and any unburnt hydrocarbon can easily be oxidised using a simple Pt catalyst. However, the NO_x remains unaffected. One solution to the control of these NO_x emissions in the oxygen-containing atmosphere of the lean-burn engine was introduced by Toyota who developed the use of NO_x traps. These traps contain BaO to adsorb the NO_x content of the exhaust gas by forming $Ba(NO_3)_2$. A schematic representation of these traps is shown in Fig. 6.20. The traps are regenerated periodically by cycling the system briefly to fuel-rich conditions in order to convert the nitrate species formed back to the original BaO.

The diesel engine also operates in the lean-burn region of but has the advantage that combustion occurs at a lower temperature than that in the Otto engine. The main gaseous emissions from a diesel engine are therefore low concentrations of unburnt fuel as well as low NO_x concentrations. However, these emissions have associated with them particulate matter that result from incomplete combustion of the fuel. The low hydrocarbon concentrations in the exhaust gases can be controlled by using a simple oxidation catalyst. However, while the particulate matter can be collected in simple filter assemblies of the type illustrated in Fig. 6.21, these filters will become gradually blocked, this resulting in deterioration of the engine performance. Regeneration of the filters can, at least in theory, be achieved reasonably simply by heating them to a sufficiently high temperature in an oxygen-containing stream, this being termed 'active regeneration'. However, such combustion is difficult to control without damaging the filter and so another method of controlling the build-up, termed 'passive regeneration', is generally used. For this, a catalyst is included within the filter (see Fig. 6.22) that oxidises any NO in the exhaust stream to NO_2. The NO_2 then catalyses the oxidation of the soot that is collected in the filter. The latter passive regeneration system operates without the need for active control involving periodic changes of the reaction conditions.

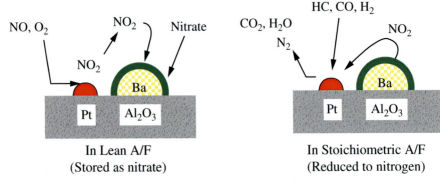

Fig. 6.20 The Toyota model for the NOx storage catalyst. *(S. Matsumoto, Catal. Today, 90 (2004) 183-190, reproduced with kind permission of Elsevier.)*

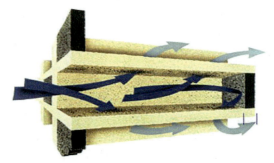

Fig. 6.21 Schematic representation of a ceramic filter for the removal of particulates from a diesel exhaust stream. *(M.V. Twigg, Catal. Today, 163 (2011) 33–41, reproduced with kind permission of Elsevier.)*

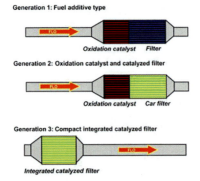

Fig. 6.22 Three different catalysed filter systems for the removal of particulates from a diesel exhaust stream. *(M.V. Twigg, Catal. Today, 163 (2011) 33–41, reproduced with kind permission of Elsevier.)*

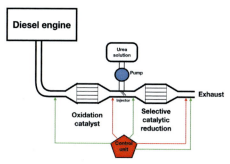

Fig. 6.23 A typical diesel selective catalytic reduction system combined with a diesel oxidation catalyst.

With increasingly strict emission regulations, there is now also a need to control the NO_x emissions from diesel engines. Under the oxidising conditions of the diesel exhaust, selective catalytic reduction using three-way catalysts as used in Otto engines cannot be used. In consequence, NO_x trapping of the type described above can be used. An alternative method of removing the NO_x is to carry out selective reduction using ammonia as reductant.[k] This process is similar to that used in the selective reduction of NO_x in the oxidising atmospheres of the exhaust gases from power stations. Fig. 6.23 shows such a selective reduction system for use in diesel systems.

The diesel exhaust gas is first passed through an oxidising catalyst to remove any unburnt hydrocarbons and CO from the stream and also to convert any NO to NO_2. The reductant is then introduced, generally as a solution of urea. The urea decomposes thermally to produce ammonia according to the equation:

$$(NH_2)_2 CO \rightarrow NH_3 + HNCO$$

The isocyanic acid formed as a byproduct is then being hydrolysed to form additional ammonia by the reaction:

$$HCNO + H_2O \rightarrow NH_3 + CO_2.$$

The reaction occurring in the selective catalytic reduction bed, which contains either a zeolite or a vanadia-containing catalyst similar to those used in the selective reduction of NOx emissions from power stations, is:

$$6\,NO_2 + 8\,NH_3 \rightarrow 7\,N_2 + 12\,H_2O$$

---

[k] See for example, 'The pollutant emissions from diesel-engine vehicles and exhaust after treatment systems', I.A. Resitogglu, K. Altinisik and A. Keskin, Clean Techn. Environ. Policy 17 (2015) 1715–27. (This extensive review is open access.)

150 Sustainable energy

The all-over performance of the selective reduction unit is monitored and controlled by a series of temperature sensors and a pair of NO_x sensors. Because of the need to have a supply of urea in order to carry out the selective reduction, such systems are most commonly found only in larger engines such as those of lorries and other heavy-duty vehicles. However, some automobile manufacturers are currently beginning to introduce similar systems into light-duty vehicles. It should be noted that many of the catalytic systems discussed above encounter problems if the fuel used contains any significant quantity of sulphur-containing impurities as sulphur is a non-reversible poison for most catalysts. The introduction of the technologies discussed above have therefore been associated with a move by the petroleum refining industry towards the production of low-sulphur containing fuels with the aim of obviating catalyst poisoning problems. The introduction of stricter and stricter regulations have however led to some serious difficulties in the control of diesel emissions that have resulted in the reduction of sales of diesel automobiles: see Box 6.3.[1]

Hybrid vehicles

With an increased awareness of the need to decrease global greenhouse gas emissions and strengthening regulations, car manufacturers started to examine methods of improving the fuel consumption behaviour of their vehicles. While many started to concentrate on the production of lean-burn petroleum- and diesel-fuelled vehicles, Toyota introduced the world's first hybrid vehicle, their Prius model, at the Tokyo Motor Show of 1995. The first production models of the Prius went on sale in December 1997 but international sales only started in 2000. Shortly after the Toyota development, Honda produced their Insight, this being launched at the Tokyo Motor Show in 1997; the Insight became generally available in Japan in 1999 and it was launched in the US in 2000, becoming the first hybrid vehicle available there.

A hybrid vehicle such as the Prius has two batteries: the 'auxiliary battery', with low voltage (12 V) which is used for starting the car and providing initial power for the on-board computer; and the second 'traction battery' (with a voltage over 250 V) of relatively low capacity which is used to power the electric traction motor. This battery is charged during braking of the car ('regenerative braking') and the generation process at the same time supplies

[1] See for example: https://en.wikipedia.org/wiki/Volkswagen_emissions_scandal.

BOX 6.3 Dieselgate

In 1999, the US introduced new Tier 2 rules decreasing over a period of several years the limit for NOx emissions from 1.0 g/mile to 0.07 g/mile. It transpired in 2015, by which time the new US limits applied, that Volkswagen (who had adopted a lean-NO_x trap approach rather than selective reduction methods) had been unable to meet the new standards for their turbocharged direct injection (TDI) diesel engines under road conditions. In consequence, they had introduced a 'defeat device' linked to the control system of the engine that allowed it to operate with acceptable limits under bench-test conditions but bypassed the system for the remainder of the time (see Fig. 6.24). The ensuing scandal caused an immediate drop in the Volkswagen share value (almost 40% drop over a period of 3 weeks) and a series of dismissals of personnel and court cases. It later transpired that very few brands of diesel engines were able to meet the Tier 2 standards under road conditions. Manufacturers affected included Volvo, Renault, Mercedes, Jeep, Hyundai, Citroen, BMW, Mazda, Fiat, Ford and Peugot. It was also shown that many brands of diesel cars failed to meet similar European road-test standards.

Fig. 6.24 A 2010 Volkswagen Golf TD1 fitted with a 'defeat device' exhibited in the Detroit Auto Show. *(https://commons.wikimedia.org/wiki/File:VW_Golf_TDI_Clean_Diesel_WAS_2010_8983.JPG.)*

Although it is possible using the selective reduction approach described able to produce diesel engines that emit very low concentrations of NOx, the attitude of the general public to diesel cars has been seriously affected. The consequence of this is that there has been a very concerted move by a number of the car manufacturers to produce battery-powered vehicles, a topic discussed later in this chapter.

152 Sustainable energy

a proportion of the braking capacity required, giving a further energy saving. The traction battery is used to power a secondary electric motor which is engaged at startup and also at lower speeds of up to 25 mph (40 km/h). The secondary motor is also used as a source of additional power for acceleration or climbing, benefitting from the higher efficiency of the electric motor compared with that of the main petrol engine. The electric motor automatically takes over as the vehicle slows down to stop and it shuts off when the vehicle is stationary, thus helping to economise on fuel consumption. The hybrid concept gives significantly improved all-over fuel economy. The Honda Insight, which operates with a primary lean-burn engine,[m] was rated as the most efficient gasoline-fuelled vehicle in the US in 2014 by having a road fuel usage of 61 miles per US gallon (73 miles per imperial gallon or 3.9 L/100 km).

The battery of a hybrid vehicle such as the Prius or Insight (or their equivalents from other manufacturers) is never completely charged, nor is it fully discharged: the range of charge is normally in the range 40%–60%; a computer-controlled operation system ensures that the level of charge never gets outside the range 38% to 82% and this effectively extends the useful life of the battery well beyond the operating age of a normal vehicle. With the Toyota Prius, the battery is of the nickel metal hydride type (see Chapter 7), made by the Panasonic EV Energy Company. If the petrol motor is not operated and the speed is kept below 25 mph, the vehicle can be driven using only the battery for very short journeys but such use would be very uncommon. It is worth noting that a number of other types of hybrid systems have also been designed, for example using compressed air as an energy source. However, such concepts are not discussed here as they do not appear to have been put into production.

Plug-in hybrid vehicles

A marked improvement on the simple hybrid design of the Toyota Prius and other equivalent vehicles such as the Honda Insight discussed above was the introduction of the 'Plug-in Hybrid' shown schematically in Fig. 6.25. Such a vehicle has a somewhat larger traction battery, normally of the Li-hydride variety (Chapter 7), and this is coupled to a relatively small electric traction

---

[m] Honda and a range of other manufacturers are now using so-called Atkinson cycle engines in their hybrid vehicles. This cycle operates in a similar way to that of the Otto engine but with a shorter compression stroke and a longer expansion stroke, this giving improved fuel economy. (See, for example: https://en.wikipedia.org/wiki/Atkinson_cycle.)

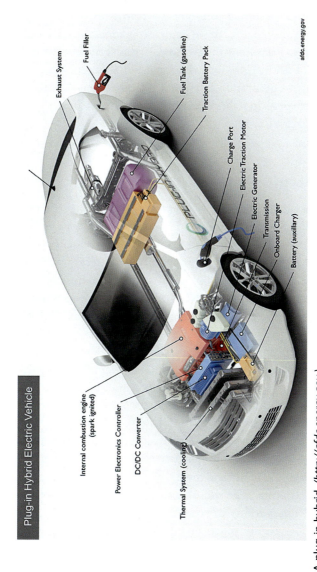

Fig. 6.25 A plug-in hybrid. *(http://afdc.energy.gov.)*

motor. The great advantage of this arrangement is that it is possible to travel significantly further when using only the stored electrical energy fed to the car from an external source. This means that it is not necessary to recharge the battery before driving further when the charge runs low since the system then changes automatically to use of the petroleum-fuelled engine. When operating with this engine, the system also gradually recharges the battery. The first Toyota Prius Plug-In Hybrid with this conformation had a range of only 18 km of normal driving without using any petroleum fuel but this figure has been significantly improved in subsequent models so that currently available plug-in hybrids do not require the use of any petroleum fuel for most local driving.[n]

Battery electrical vehicles

Automobiles powered solely by batteries are very recent developments although battery power was used in specialised vehicles such as milk floats and fork-lift trucks during the previous century, as discussed earlier in the chapter. The sudden rise in the popularity of battery-powered cars in the last 15–20 years is such that the global share of such vehicles is expected to rise to more than 20% by 2030, having only been 2% as recently as 2016. Some countries have announced that they intend to ban the sale of petroleum-powered vehicles in the coming years and many have introduced incentives for the purchase of battery-driven cars. The increased importance of electric vehicles is associated with the very rapid developments of relatively inexpensive and reliable batteries such as those that are currently used in the plug-in hybrid vehicles discussed above. These improvements have been coupled with rapidly increasing battery storage capacities and corresponding increases in driving distances before recharging is necessary. Also contributing to the surge in popularity has been the wide-ranging installation of recharging facilities and the development of equipment to give much more rapid charging. The use of modern rechargeable batteries has been extended to other forms of road transport such as trucks, buses, trains, scooters and bicycles as well as aircraft. For a full discussion of the construction and operation of such batteries, see Chapter 7.

---

[n] Another variant that does not appear to have operated until now is a battery powered vehicle with a small low-powered petrol engine that is used only to recharge the battery as the electrical charge is used up. If such an engine was fuelled with a biomass-derived fuel and the electricity for the battery was provided from totally renewable sources, this type of vehicle would be completely 'zero emission'.

A number of companies manufacture all-electric vehicles. The best-selling models come from Tesla and Nissan: in March 2020, 500,000 units of the Tesla Model 3 had been delivered and the Nissan leaf passed the 500,000 mark in December 2020. Other successful manufacturers include Hyundai, Jaguar, Kia, a Renault-Nissan-Mitsubishi alliance, BMW and Volkswagen. Electric cars are also manufactured in China by the BAIC Group (Beijing Automotive Industry Holding Company Ltd.), SAIC-GM-Wuling Automobile (a joint venture including General Motors) and Chery (Chery Automobile Co. Ltd.). Fig. 6.26 shows the layout of such a vehicle while Fig. 6.27 is a photograph of the chassis of a Tesla model (Model S) showing the traction engine (nearest camera) and the battery compartment. There has been a very steady increase in the range of battery-powered electrical vehicles. Table 6.1 lists the ranges of some of the models that were available commercially in the UK in early 2021. The Tesla models have the highest ranges in the listing, with values that are comparable with those of the most efficient petrol cars. However, the other lower priced models still have significant ranges that begin to make them very competitive with internal combustion engines. Short recharging times are also important and Tesla has been at the forefront in introducing rapid charging stations for its models, particularly in California; with the fast-charging units, it is claimed that the batteries will reach 80% power in 30 min. Nevertheless, charging times for most models are anything between 4 and 8 h, these values being quoted for the BMWi3 and the Nissan Leaf.[o]

Another very important factor in relation to the use of electric vehicles is whether or not the electricity used for charging the vehicle is from a fully renewable source or not. For this reason, countries or regions in which the proportion of renewable electricity is high are more favourable locations for the use of such vehicles. A recently published report[p] shows that if only 50% of the electricity available in the grid system is renewable, the use of plug-in hybrid vehicles gives a greater reduction in CO_2 emissions than does the use of fully electrical vehicles. Table 6.2 summarises some of these conclusions; it should be noted that these data also take into account the

---

[o] These data were taken from the following site: https://youmatter.world/en/hydrogen-electric-cars-sustainability-28156/ but similar results are to be found on other equivalent sites. The 'youmatter.world' site is also a useful source of additional material relevant to the content of this book.

[p] 'The European Environment - State and Outlook 2020', European Environment Agency (2019), ISBN 978-92-9480-090-9. The full report is downloadable from http://europa.eu; https://www.forbes.com/sites/davidrvetter/2021/01/25/its-official-in-2020-renewable-energy-beat-fossil-fuels-across-europe/?sh=fa3872d60e83.

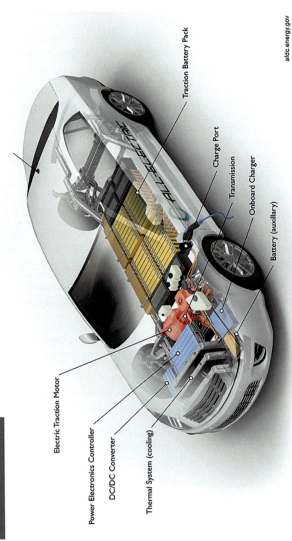

Fig. 6.26 Schematic representation of an electric vehicle. (*http://afdc.energy.gov.*)

Fig. 6.27 A Tesla Motors Model S base. *(https://upload.wikimedia.org/wikipedia/commons/f/f3/Tesla_Motors_Model_S_base.JPG)*

Table 6.1 A comparison of the ranges of some of the models of electric cars now available showing the top ten best ranges.

Manufacturer/Model	Maximum range/miles	Price/£
Tesla Model S Long Range	379	77,980
Tesla Model 3 Long Range	348	46,990
Tesla Model X Long Range	314	82,980
Jaguar i-Pace	292	64,495
Kia e-Niro	282	36,495
Hyundai Kona Electric	278	3890
Mercedes-Benz EQC	259	65,720
Audi e-tron	239	58,900
Nissan Leaf e$^+$	239	35,895
BMW i3	193	37,480

(Source of data: Car Magazine, 4 January 2021. https://www.carmagazine.co.uk/electric/longest-range-electric-cars-ev/.)

green-house gas emissions associated with the production of both the fuel used and the vehicle as well as that expected for its eventual recycling at end of life. Only if the vehicle is used in regions where the percentage of renewable electricity production exceeds the average value for the EU (about 58% in 2020[q]) does a fully electric vehicle offer an advantage over the plug-in hybrid vehicle when all the factors regarding its life-time operation are taken into account. As an example of a country where the percentage of renewable electricity is high is Norway which produces 98% of its

---

[q] https://www.forbes.com/sites/davidrvetter/2021/01/25/its-official-in-2020-renewable-energy-beat-fossil-fuels-across-europe/?sh=fa3872d60e83.

Table 6.2 Life-cycle emissions of CO_2 for a series of different vehicles and fuel types.

Vehicle type	Vehicle production and disposal/CO_2 emissions (g/km)	Fuel production / CO_2 emissions (g/km)	CO_2 exhaust emission/ CO_2 emissions (g/km)	Total life-cycle CO_2 emissions /(g/km)
Petrol	40	25	170	235
Diesel	40	30	130	200
Plug-in Hybrid	55	25	90	170
Battery Electric/100% renewable	60	20	0	80
Battery Electric/EU Average Renewable Electricity	60	115	0	175
Battery Electric/100% Coal Generation	60	240	0	300

(Data from 'The European environment - state and outlook 2020', European Environment Agency (2019), ISBN 978-92-9480-090-9 (http://europa.eu).)

Transport **159**

electricity using hydropower; as a result, the proportion of new electric vehicles there is very high, reaching 54% of the total sales in 2020.

Fuel cell vehicles

As discussed above, the relatively long recharging time of an electric vehicle is a drawback in many instances since a longer journey has to be well planned to allow relatively long stop-overs to achieve reasonable levels of recharge. An alternative concept, now being marketed in California and in a limited number of other locations, is the fuel-cell vehicle. For these, the fuel is hydrogen gas and the range of a typical model is roughly equivalent to that of a conventional internal combustion engine powered vehicle. The topic of fuel cells is discussed more fully in Chapter 7. With pure hydrogen as fuel,[r] the fuel cell has the great advantage that it produces only water as product and so the problem of green-house gas creation relates back to the method of production of the hydrogen: whether or not it is 'green'. Important factors of relevance to the operation of fuel cell vehicles are therefore the availability of green hydrogen and its cost. As shown in Fig. 6.28, very few countries or regions have yet established a suitable infrastructure for the widespread sale of hydrogen and relatively few filling stations have been created; only a proportion of these are available for public use, the remainder being for the use of private fleet vehicles. The figure also shows the planned introduction of further refuelling stations, the majority of the future installations being in the EU and Asia (especially China and Korea).

Fig. 6.29 shows the layout of a typical fuel cell automobile. The fuel cell stack is relatively small compared with the battery storage compartment shown in Fig. 6.26. The hydrogen used as fuel is stored as a gas at a pressure of about 700 atm (70 MPa) and this requires an especially strong fuel tank. The tank used in the Toyota Mirai, one of the currently available commercial vehicles, has a three-layer structure made of plastic that is reinforced by carbon fibre and various 'other materials'. Toyota say that they have introduced a high-capacity converter to boost the voltage of the system to 650 V, thereby enabling a decrease of the size of the fuel cell stack compared with a previous model that had previously been available for leasing only in California.[s] The advantage of the fuel-cell vehicles compared with battery

---

[r] As discussed in this chapter, fuel cells can also be operated using hydrogen generated in-situ from a fuel such as methanol. Although prototypes of a wide variety of vehicles using methanol fuelled fuel cells have been produced and operated, they are not in general operation.

[s] https://www.toyota.ie/world-of-toyota/articles-news-events/2014/mirai-fuel-cell.json.

160 Sustainable energy

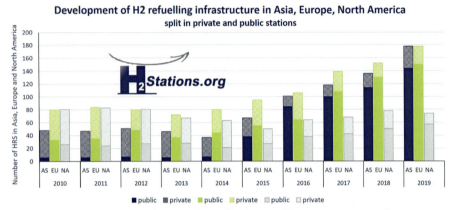

Fig. 6.28 The development of the numbers of hydrogen filling stations in Asia, the European Union and North America from 2011 to 2019. *(www.H2stations.org. Reproduced with kind permission of Ludwig-Bölkow-Systemtechnik GmbH (LBST).)*

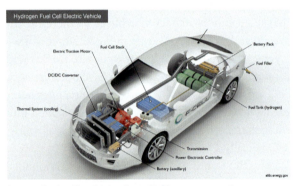

Fig. 6.29 Hydrogen fuel cell vehicle. *(http://afdc.energy.gov.)*

electric vehicles is that the tank can be refilled in less than 5 min and the range is of the order of 300 miles, a value comparable to that of a petroleum-fuelled vehicle.

Concluding remarks

This chapter has shown the development of different means of transport, with particular attention being given to the development of the personal automobile and of low-emission technologies used for them. In most situations it would appear that plug-in hybrid cars are currently among the most environmentally friendly in terms of both contaminants and CO_2 emissions although all-electric battery vehicles win out in regions where a significant proportion of the electric supply is produced from renewable sources. Battery vehicles will also be preferable if the home-owner is equipped to generate the required electricity, for example by using a photovoltaic system, and if the potential range of the vehicle between recharging is adequate for the owners requirements. Despite much publicity regarding the use of hydrogen as a fuel for automobiles and the relatively long history of the development of fuel cells suitable for such use, the lack of an adequate distribution network for hydrogen is a major drawback to the acceptance of hydrogen powered vehicles. Sales of fuel-cell powered cars are therefore likely to be restricted to the very small number of locations where public hydrogen-fuelling stations are situated. Hydrogen is more likely to be used as a fuel in larger vehicles operated from a privately operated hydrogen supply depot and it is probable that there will be significant developments of such usage. Batteries and fuel cells are also very important in relation to energy generation and also to the storage of electrical power. The following chapter (Chapter 7) deals in more detail with fundamental aspects of the development of batteries and fuel cells and discusses their uses for transportation and other purposes.

CHAPTER 7

Batteries, fuel cells and electrolysis

Introduction

With the increasing availability of renewable electrical energy from a wide variety of sources such as wind and photovoltaic arrays, there is an increasing demand for methods of storing this energy in such a way that excess generated in periods of peak production can be stored and then used at other times. Both batteries and fuel cells have important contributions to make in such storage and usage strategies. This chapter addresses some aspects of these two related technologies. It starts with a brief history of their discoveries and developments, proceeds to an outline of some of the background electrochemical science of various battery systems and fuel cell devices, and finally discusses some related electrolysis processes. As some of the uses of batteries and fuel cells in transport applications, particularly in automobiles, were discussed in Chapter 6, the focus here is on other applications.

The Volta pile, Faraday and the electrochemical series

The Italian Physicist, Alessandro Volta, was the first to demonstrate (in 1799) that an electric current could be produced in an external circuit when alternating discs of copper and zinc were placed one on top of another, each layer being separated by either a cloth or a piece of cardboard soaked in brine. The structure of this 'voltaic pile' is illustrated in Fig. 7.1. Volta believed that this current was due to a voltage difference between the metals but Humphry Davy showed that the electromotive force (EMF) that was produced was due to the occurrence of a chemical reaction. When a current flows through the cell, metallic zinc at the surface of a zinc layer (the 'anode') is oxidised to Zn^{2+} and two free electrons are liberated into the metal:

$$Zn \rightarrow Zn^{2+} + 2\,e^-$$

Sustainable Energy
https://doi.org/10.1016/B978-0-12-823375-7.00003-2

Copyright © 2022 Elsevier B.V.
All rights reserved.

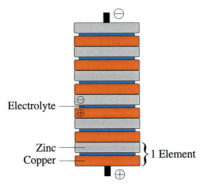

Fig. 7.1 A voltaic pile. *(https://upload.wikimedia.org/wikipedia/commons/0/06/Voltaic_pile.svg.)*

These electrons pass through the external circuit to the copper ('cathode') where they react at the surface with H+ ions from the brine 'electrolyte':

$$2H + 2\,e^- \rightarrow H_2$$

The allover reaction can thus be written as:

$$Zn + 2H^+ \rightarrow Zn^{2+} + H_2$$

When no current is drawn from the pile, each cell of the stack generates a voltage of 0.76 V with a brine electrolyte. With the stacking shown in Fig. 7.1, six cells in series, the total EMF produced by the stack is $0.76 \times 6 = 4.56$ V.

In this example, the copper of the stack does not take part in the allover chemical reaction but only acts as a catalyst for the proton transfer reaction since hydrogen is evolved at the Cu electrode which remains inert. It was subsequently shown that many other materials could be used as the cathode as long as they are inert under these conditions, examples being silver, platinum, stainless steel or graphite. If a current continues to be drawn, the zinc electrodes are completely consumed as long as sufficient electrolyte is present to provide the H^+ ions needed for the reaction and the Zn^{2+} ions can remain in solution without fouling up the electrodes. In principle, it would also be possible to reverse the reaction as long as gaseous hydrogen was provided to the copper electrode compartment and a suitable voltage of greater than 0.76 V was applied to the cell in question.

Volta's work caused a great deal of excitement in the scientific community and many well-known scientists of the period became involved in carrying out research on related topics. One of these was Michael Faraday, then working as an assistant to Humphry Davy at the Royal Institution in London, who showed that all types of electricity that had been studied up until that time, including voltaic, thermal and magnetic, were equivalent. Faraday went on to carry out experiments on many aspects of electricity, introducing in 1834 the term *electrolysis* and the laws governing the process. In his lectures given to the Royal Institution, he also popularised the use of what are now familiar names: anode, cathode, electrode and ion.

Faraday's Laws of Electrolysis recognised that there is a direct relationship between the voltage generated in any electrochemical cell (such as the Zn-Cu cell of the Volta pile) and the chemistry of the oxidation of the metal involved (the Zn in the Volta pile). Simply stated, the laws are as follows:

First Law. The mass of an element deposited at an electrode (m) is directly proportional to the charge, Q, passed through the electrode (in ampere seconds or Coulombs (C)):

$$m/Q = Z$$

where Z is the electrochemical equivalent of the substance involved.

Second Law. The mass of any substance liberated or deposited at an electrode is directly proportional to their chemical equivalent weight (W):

$$W = \text{Molar mass}/\text{Valence}$$

As long as the electrochemical process in question is carried out reversibly at a constant temperature and pressure, the consequence of these relationships is that the EMF of any cell reaction, E, is directly related to the free energy change associated with that reaction (ΔG) by the equation:

$$\Delta G = -nFE$$

where n is the number of electrons required per ion reacted and F is the Faraday constant. (The value of F is $96,485.332 \, C \, mol^{-1}$, this generally being rounded to $96,485 \, C \, mol^{-1}$.) Hence, the all-over EMF of any combination of electrodes measured under reversible conditions is directly related to the chemical thermodynamics of the chemical reaction concerned. For a more detailed discussion of the thermodynamics of electrochemical reactions see Box 7.1.

BOX 7.1 The thermodynamics of a reaction in an electrochemical cell

In order to discuss the operation of modern batteries, fuel cells and related systems, it is necessary first to consider briefly some thermodynamic and kinetic aspects of the operation of electrochemical cells. For any reaction:

$$aA + bB \Leftrightarrow cC + dD$$

the value of the free energy change ΔG is given by Eq. (7.1):

$$\Delta G = \Delta G^\circ + RT \ln\left(aC^c.aD^d/aA^a.aB^b\right) \tag{7.1}$$

where ΔG° is the standard free energy change for the all-over reaction. If the reaction is at equilibrium, ΔG is zero and

$$\Delta G^\circ = -RT \ln K \tag{7.2}$$

where K is the equilibrium constant for the reaction. The quantity ΔG in Eq. (7.1) is an indication of how far the reaction is from equilibrium; a positive value corresponds to composition to the left of the equilibrium position, a negative value to composition to the right. If a random mixture of A, B, C and D is made up and there are no constraints to the reaction occurring, the value of ΔG will tend to zero and an equilibrium composition will be attained. As discussed above, the free energy change for any electrochemical half-cell reaction is given by:

$$\Delta G = -nFE \tag{7.3}$$

Substituting the values of ΔG and ΔG° in Eq. (7.1) by $-nFE$ and $-nFE^\circ$, where n is again the number of electrons transferred in the reaction and E° is the standard EMF of the cell, we obtain:

$$nFE = nFE^\circ - RT \ln\left(aC^c.aD^d/aA^a.aB^b\right)$$

or

$$E = E^\circ - RT/nF. \ln\left(aC^c.aD^d/aA^a.aB^b\right) \tag{7.4}$$

It should be recognised that E° is therefore directly related to the standard free energy of the reaction occurring and the equilibrium constant for the reaction, K, by the following relationship:

$$E^\circ = -\Delta G^\circ/nF = +RT\ln K/nF \tag{7.5}$$

While a normal chemical reaction will proceed spontaneously towards its equilibrium position with the corresponding value of ΔG tending to zero, the position of an electrochemical reaction can be controlled by the value of the potential E applied to the electrodes. In other words, it is possible by the application of a potential to bring about a reaction that is not thermodynamically permitted in the absence of this electrochemical potential. A battery is an example of a combination of half-cell reactions involving an allover spontaneous chemical reaction. In contrast, electrolysis is an example of a reaction which would not be possible without the application of a potential.

Half-cell EMF's and the electrochemical series

An important consequence of the recognition by Faraday that the EMF associated with an electrochemical process is thermodynamically controlled and that the EMF of any electrochemical process can be determined from the thermodynamics of that process led to the establishment of the so-called 'electrochemical series'. This gives the standard half-cell EMF of any single electrode process relative to that of a standard hydrogen half-cell electrode of the type shown in Fig. 7.2.

This standard hydrogen half-cell involves the reaction of hydrogen ions at an inert electrode composed of a metal such as platinum immersed in a molar acid solution at 298 K and with a hydrogen pressure of 1 atm. (101.325 Pa, 1.01325 bar):

$$2H^+ + 2e^- \Leftrightarrow H_2$$

Table 7.1 gives some selected values of standard half-cell potentials measured under reversible conditions at 298 K.[a] Each value given corresponds to the potential related to a single electron transfer. All species are at concentrations of 1 mol/L, these corresponding to activities of unity for each pure solid, pure liquid or water (when used as a solvent).

As an example of the use of the data in Table 7.1, consider the Zn-Cu cell of the Volta pile shown in Fig. 7.1. The half-cell potential of the Zn electrode ($Zn^{2+} \to Zn$) is given as -0.7618 V while that of the Cu electrode ($Cu^{2+} \to Cu$) is given as $+0.337$ V. The reaction at the Zn electrode will

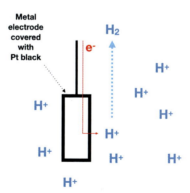

Fig.7.2 A hydrogen electrode consists of a metal coated with Pt black and immersed in an acidic solution. Electrons from the external circuit react with hydrogen ions from the solution to form hydrogen gas.

---

[a] A more complete listing of the electrics-chemical series is given at: https://en.wikipedia.org/wiki/Standard_electrode_potential_(data_page). This listing includes values for many other metals as well as for some commonly encountered chemical compounds that can be used in electrode reactions.

168 Sustainable energy

Table 7.1 Standard electrode potentials.

Element	Reactant	No. of electrons involved in the reaction	Product	$E°$/Volts
Li	Li^+	1	Li(s)	−3.040
K	K^+	1	K(s)	−2.931
Ba	Ba^{2+}	2	Ba(s)	−2.912
Sr	Sr^{2+}	2	Sr(s)	−2.899
Ca	Ca^{2+}	2	Ca(s)	−2.868
Na	Na^+	1	Na(s)	−2.71
Mg	Mg^{2+}	2	Mg(s)	−2.70
La	La^{3+}	3	La(s)	−2.379
Y	Y^{3+}	3	Y(s)	−2.372
Mg	Mg^{2+}	2	Mg(s)	−2.372
Ce	Ce^{3+}	3	Ce(s)	−2.336
Sr	Sr^{2+}	2	Sr(s)	−1.793
Al	Al^{3+}	3	Al(s)	−1.662
Zr	Zr^{4+}	4	Zr(s)	−1.45
H	$2 H_2O$	2	$H_2 + 2 OH^-$	−0.828
Zn	Zn^{2+}	2	Zn(s)	−0.7618
Cr	Cr^{3+}	3	Cr(s)	−0.74
Ta	Ta^{3+}	3	Ta(s)	−0,6
Co	Co^{2+}	2	Co(s)	−0.28
Ni	Ni^{2+}	2	Ni(s)	−0.25
C	$CO_2(g) + 2H^+$	2	$CO_2 + H_2O$	−0.11
C	$CO_2(g) + 2H^+$	2	HCOOH (aq)	−0.11
H	$2H^+$	2	H_2	0.00
Cu	Cu^{2+}	1	Cu^+	+0.159
Cu	Cu^{2+}	2	Cu(s)	+0.337
Cu	Cu^+	1	Cu(s)	+0.520
O_2	$O_2(g) + 2H_2O(l)$	4	$4 OH^-$	+0.401
Ag	Ag^+	1	Ag(s)	+0.7996
Hg	$Hg2^{2+}$	2	2Hg(l)	+0.80
Pd	Pd^{2+}	2	Pd(s)	+0.915
Pt	Pt^{2+}	2	Pt(s)	+1.188
Cl_2	$Cl_2(g)$	2	$2Cl^-$	+1.36
Au	Au^+	1	Au(s)	+1.83

thus proceed through the oxidation of Zn metal to Zn^{2+} ions in solution. However, as the half-cell potential of the Cu electrode is positive and thus above the value for the hydrogen electrode, hydrogen evolution according to $2H^+ \rightarrow H_2$ occurs at the Cu electrode; even if Cu^{2+} ions had been available in solution so that Cu deposition might have occurred, the hydrogen evolution reaction is thermodynamically preferred. It is possible to construct a large number of different electrochemical cells using various combinations of half cells such as those listed in Table 7.1. These can consist simply of a container in which a single electrolyte exists between the two electrodes, as in the Volta pile discussed above. However, more commonly, they involve two separate half-cell assemblies linked to one another in such a way that a current can pass between the two electrodes. Fig. 7.3 illustrates one such combination in which the anode and cathode compartments are separated by a salt bridge. (A salt bridge is a tube that typically contains a highly concentrated solution of KNO_3—often maintained in an agar gel—the ions of which do not react with either the anode or cathode; see Box 7.2.) The charge balance of the system is then maintained by the movement of K^+ ions to the cathode compartment to compensate for the creation thereof cations by the reaction of the cathode material. In parallel to that, the Cl^- ions perform a similar function in the anode compartment by compensating for the removal of anions at the anode. As an alternative to the salt bridge, the two compartments can be separated by a porous membrane which allows the transport of cations and anions between the two cells but without significant mixing of the solutions in the separate

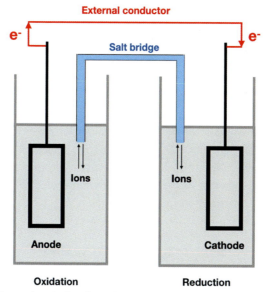

Fig. 7.3 An electrochemical cell with salt bridge. A typical electrochemical cell construction with a salt bridge separating the anode and cathode compartments. In the case of a Daniel cell (see below), the anode is Zn and this is immersed in a solution of $ZnSO_4$ while the cathode is Cu immersed in a solution of $CuSO_4$. The salt bridge can be replaced by a membrane separator.

> ## BOX 7.2 The salt bridge
>
> A salt bridge such as that which can be used in a Daniell cell is a tube that typically contains a highly concentrated solution of KNO_3, often held in an agar gel, the ions of which do not react with either the anode or the cathode. The charge balance of the system is then maintained by the movement of K^+ ions to the cathode compartment to compensate for the removal of Cu^{2+} ions and of NO_3^- ions to the anode compartment to balance the charge of the liberated Zn^{2+} ions. As with the use of a porous membrane, the cell gradually loses potential when used in either a continuous battery mode or a recycling mode due to the mixing of the solutions on either side of the salt bridge.

compartments; see Box 7.2. The Volta pile discussed above functions as a battery, producing electricity until the components responsible for the current are completely used up. The reaction is not reversible as the hydrogen produced in the reaction is released irreversibly to the atmosphere. However, many combinations of half-cell electrodes operate reversibly since the chemicals involved are retained in the containing vessel. Hence, they can in principle be used either for the provision of electrical energy (operating as a battery) or for the storage of electricity supplied to the combination from an external source. Electrolysis is a related process which also involves two half-cell electrodes; the most common example, discussed more fully later in the chapter, is the production of hydrogen and oxygen by the electrolysis of water.

Another important related development that occurred sometime after the work of Volta was the invention of the fuel cell by Sir William Grove in 1838, a topic that is also discussed more fully in a later section of this chapter. The fuel cell involves the production of an electric current by the reaction of suitable chemicals, most frequently hydrogen and oxygen (the reaction being the reverse of the example given for electrolysis), these gases being provided externally to the system.

The kinetics of electrochemical processes

Just as a process occurring at an electrode is thermodynamically controlled, once the conditions deviate from equilibrium, any reaction occurring is also subject to kinetic control. As discussed above, all electrode processes involve electron transfer and the electrons created to travel through an external circuit, the current resulting being a measure of the rate of the electrode process. The current produced in any half-cell reaction is determined by two factors, the *Exchange Current Density*, j_o, for the electrode and the *Overpotential*, η, applied to that electrode. Just as when a chemical reaction is at equilibrium, there is a dynamic situation in which both forward and reverse reactions continue to occur with the forward and reverse reaction rates being equal, an electrochemical reaction is a dynamic situation in which ions are formed and discharged at equal rates. This rate is determined by the *Exchange Current Density*, j_o. Only when an *Overpotential* is applied is there a net current flow to or from the electrode. As the current flowing through each half-cell electrode in an operating cell combination is the same, the overpotentials at the two electrodes will be different. Box 7.3 gives more detail of the theory relating to exchange current density and overpotential, showing the importance of the electrical *double layer* existing at the electrode surface.

BOX 7.3 Exchange current density and overpotential

Consider again the reaction:

$$aA + bB \rightarrow cC + dD$$

but consider now that the rate of this reaction is now kinetically controlled. The conversion rate will now be given by a kinetic expression such as:

$$\text{Rate} = -d[A]/dt = -d[B]/dt = f(A, B).\exp(-E_a/RT)$$

where $f(A, B)$ is some function of the concentrations of A and B and the exponential term includes the 'activation energy' of the reaction, E_a. The all-over reaction may occur in a complex sequence of reaction steps but there is often a single rate-determining step. In an electrochemical process such as that occurring in an electrochemical cell, the rate of that process is in most cases determined by the rate of transfer of ionic species in the electrolyte towards the electrode involved and the all-over reaction rate can be determined by the rate of reaction at the anode or the cathode (or even a combination of both). It is well established that a so-called *double layer* is established at an electrode as depicted in Fig. 7.4.

This model (a slightly refined model of the Gouy-Chapman model due to Bokris, Devanthan and Müllen) shows the surface of the negatively charged cathode as cations approach it. (An equivalent diagram would apply to the movement of anions towards the anode.) The cations first pass through layer 3, the 'Diffuse Boundary Layer', then through a layer containing well-ordered hydrated cations, the 'Outer Helmholtz plane', and finally they encounter the innermost layer of adsorbed solvent molecules, the 'Inner Helmholtz plane'. The double-layer

Continued

BOX 7.3 Exchange current density and overpotential—cont'd

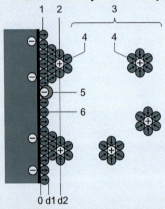

Fig. 7.4 Schematic representation of a double layer. (Bokris/Devanathan/Müllen model.) 1. Inner Helmholtz plane; 2. Outer HP; 3. Diffuse layer; 4. solvated ions (cations); 5. specifically adsorbed ions (redox ion contributes to pseudocapacitance); 6. molecules of electrolyte solvent. *(https://upload.wikimedia.org/wikipedia/commons/7/7e/Electric_double-layer_%28BMD_model%29_NT.PNG)*

provides resistance to the movement of the cations towards the cathode and the current density, *j*, is given by the Butler-Volmer equation:

$$j = j_o\{\exp(1-\alpha)F\eta/RT - \exp \alpha F\eta/RT\}$$

The quantity η is termed the 'over-potential' and is equal to ($E'-E$), where E is the electrode potential at equilibrium and E' is the electrode potential when a current is being drawn from the cell. When the over-potential is zero, the reaction at the cathode is at equilibrium and the current density is j_o, this being equal to the rate of transfer of ions through the double layer in both directions. The term α is the so-called 'transfer coefficient' and is related to the position of the electron-transfer occurrence within the double layer. The dependence of the ratio j/j_o on the over-potential for different values of the transfer coefficient is shown in Fig. 7.5.

In practice, the value of α is in most cases very close to 0.5. When the over-potential is very small (in practice ca. 0.01 V), the double layer acts like a normal conductor in which the current density is proportional to the applied voltage. However, at higher values of overpotential, there is a logarithmic relationship between the current density and the overpotential that is given by the so-called Tafel equation:

$$\ln j = \ln j_o - aF\eta/RT$$

BOX 7.3 Exchange current density and overpotential—cont'd

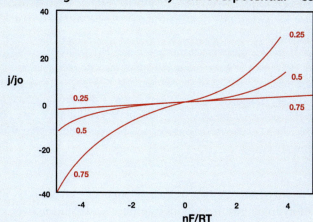

Fig. 7.5 The variation of j/j_o with overpotential for different values of the transfer coefficient.

A plot of the logarithm of current density j versus the overpotential η gives results of the type shown schematically in Fig. 7.6. The intercept on the log j at zero overpotential gives the value of the logarithm of the exchange current density and the gradient gives the value of the transfer coefficient, α, in this case 0.58.

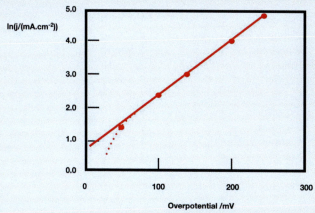

Fig. 7.6 A typical Tafel plot of versus over-potential, η.

In practice, all electrochemical devices when in operation either provide or draw a current. For example, when a rechargeable battery is being charged, a

Continued

174 Sustainable energy

> ## BOX 7.3 Exchange current density and overpotential—cont'd
> voltage must be applied that is greater than the standard EMF of the electrode assembly in question. However, the magnitude of this overpotential must not be too high or there is a danger that unwanted reactions will occur which shorten the battery life. Hence, it is very important that the rate of recharge is carefully controlled to minimise the possibility of damage. Similarly, when a battery is in use and a current is being drawn from it, the rate of discharge must be carefully controlled, this also avoiding damage to the battery. These topics are discussed further below in relation to the construction and operation of several different electrochemical devices.

Table 7.2 shows some values of the exchange current density, jo, for the evolution of hydrogen from aqueous sulphuric acid solution over a series of transition metals at room temperature. It can be seen that there is a very large range of values, the highest being for the noble metals, palladium, platinum, rhodium and iridium and the lowest being for cadmium, manganese, lead and mercury.[b]

Table 7.2 Exchange current densities for various metals for the hydrogen evolution reaction in an aqueous 1.0 N H_2SO_4 solution at ambient temperature.

Metal	$j_o/\text{amp cm}^{-2}$
Palladium	1.0×10^{-3}
Platinum	8.0×10^{-4}
Rhodium	2.5×10^{-4}
Iridium	2.0×10^{-4}
Nickel	7.0×10^{-6}
Gold	4.0×10^{-6}
Tungsten	1.3×10^{-6}
Titanium	7.0×10^{-8}
Cadmium	1.5×10^{-11}
Manganese	1.3×10^{-11}
Lead	1.0×10^{-12}
Mercury	0.5×10^{-13}

Data from several standard tabulations.

---

[b] As will be seen later, the high exchange current density for this reaction over noble metals such as platinum is of importance for the construction of electrolytic cells used for the production of hydrogen. The surface area of the platinum is optimised in such systems by coating the electrode with platinum black, a highly dispersed form of the metal.

Electrochemical batteries

The Daniel Cell. The various types of electrochemical devices that have been developed since the early work of Volta, Faraday and their contemporaries will now be discussed, starting with a description of a variety of types of battery. The first example of a cell construction in which hydrogen evolution does not occur and which demonstrates the use of the data of Table 7.1 is the so-called Daniell cell. This cell, invented in 1836 by the British scientist, John Frederic Daniell (see Fig. 7.7), uses the same Cu and Zn electrodes as in the Volta pile but these are positioned in two separate vessels connected by a membrane barrier to allow the passage of ions between the two compartments; a salt bridge can also be used (See Box 7.2 and Fig. 7.3).

The anode is Zn metal and this is immersed in a solution of zinc sulphate; the cathode is Cu and is immersed in a solution of copper sulphate. When the cell is operated as a battery, Zn^{2+} ions are liberated at the anode surface

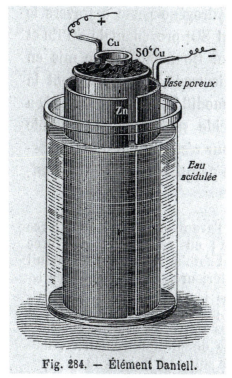

Fig. 284. — Élément Daniell.

Fig. 7.7 An historical diagram of the Daniell cell from 1904. *(https://en.wikipedia.org/wiki/Daniell_cell.).*

while Cu^{2+} ions in the cathode compartment are simultaneously reduced to Cu metal that deposits on the cathode. The porous membrane allows the transport of Zn^{2+} and $SO_4^=$ ions between the two cells but without significant mixing of the sulphate solutions, thus maintaining charge neutrality in the two vessels. The total voltage created by the combined reactions, the so-called 'open-cell voltage', is the sum of the two half-cell potentials, $0.7618 + 0.340 = 1.1018 V$, and a current is generated which passes through the external circuit. The allover reaction is:

$$Zn_{(s)} + Cu^{2+}{}_{(aq)} \rightarrow Zn^{2+}{}_{(aq)} + Cu_{(s)}$$

The zinc electrode is thus corroded during operation while the copper electrode accumulates extra copper. The reaction can be reversed if a suitable potential source is applied across the cell so that the voltage is greater than $1.1 V$. As a result, the Daniel cell can also be used as a storage device. The EMF of the Daniell cell gradually drops off with time during use due to the deposition of copper in the pores of the membrane barrier and also because of the gradual mixing of Cu and Zn ions between the two vessels.

When the Daniel cell is operated to provide power, acting as a battery, the reaction will continue until all the Cu^{2+} ions have been reduced. The actual voltage created will be less than the open-cell voltage due to the over-voltages required to give the necessary rate of reaction and due to factors such as the internal resistance of the assembly and of the external circuit; the voltage will also gradually decrease as the concentration of the Cu^{2+} solution is depleted. As the allover reaction between metallic zinc and Cu^{2+} ions is exothermic, the temperature of the cell increases, this also causing a decrease in cell voltage.

The Lead Acid Battery. The lead-acid battery will be familiar to the reader as being an important component of the majority of all motor vehicles currently in use, supplying the power for startup and for the operation of on-board instrumentation(Chapter 6). It was first invented by Gaston Planté (a French physicist) in 1859 and it was the first practical example of a truly rechargeable battery. Both the electrodes are composed of lead in the form of a lead-alloy grid. The cathode is coated with sponge lead and the anode is coated with lead dioxide (this having metallic conductivity), both being immersed in a solution of sulphuric acid. The reaction occurring at the negative plate (cathode) is:

$$Pb(s) + HSO_4^-{}_{(aq)} \rightarrow PbSO_{4(s)} + H^+{}_{(aq)} + 2e^-$$

and the reaction at the positive plate (anode) is:

$$PbO_{2(s)} + HSO_4^{-}{}_{(aq)} + 3H^+ + 2e^- \rightarrow 2PbSO_{4(s)} + 2H_2O_{(l)}$$

The total reaction is thus:

$$Pb_{(s)} + PbO_{2(s)} + 2H_2SO_{4(aq)} \rightarrow 2PbSO_{4(s)} + 2H_2O_{(l)}$$

The total standard cell voltage, $E_{cell} = 2.05\,V$.

When the battery is fully discharged, both electrodes have become coated with lead sulphate and the sulphuric acid contained in the volume between them is significantly more dilute than with the fully charged battery as a result of the formation of water. When the battery is recharged, this water is used up and the strength of the acid is again increased; at the same time, the positive plate regains its coating of lead dioxide and the negative plate becomes again pure lead. If the recharging process is carried out too quickly, the overvoltage increases to a level at which hydrogen can be formed by electrolysis and the total water content of the battery is thus depleted. Because of this possibility of hydrogen evolution, thus creating an explosive hazard, the recharging process has to be carefully controlled in such a way that the rate is not too high.

Most lead-acid batteries have either six or twelve cells in series, giving output potentials of 12 or 24 V. Apart from their use in automobiles, where their ability to supply the large currents required for the start-up ignition, they are used for backup power supplies. They are also used as power supplies for the electrical motors of traditional submarines for which the use of internal combustion engines would be undesirable. For such uses, it is important to be able to measure the level of charge of the bank of batteries. This is often done for the batteries by measuring the specific gravity of the sulphuric acid, this being directly related to its concentration. Lead-acid batteries have a relatively low electrical capacity as a function of their weight and so it is much more preferable for many applications to use one of the more modern types of battery such as the Li-ion battery to be discussed below. Lead-acid batteries generally have reasonable lifetimes during which they can be recharged many times. However, these batteries do age, particularly if left unused at a low level of charge, when a process known as sulphation occurs; although lead sulphate in a finely divided form is a participant in the reversible process that produces an electric current (see equations above), crystallisation of the sulphate occurs when the battery is left uncharged for a long period. This crystalline fraction is then no longer able to participate in the charging/discharging cycle, the result being a decrease in the power output of the battery to such an extent that it cannot any longer be used. Fortunately, however, it is relatively easy to recycle used lead-acid batteries and the process used for such recycling is generally very efficient; for example, it has been reported that 99% of the lead from used batteries in the USA is recycled. The size of the industry involved in lead-acid battery manufacture and recycling can be seen from the total worldwide use of lead in batteries. This has been estimated as being more than 1,00,000 metric tons per year.

Dry Cell Batteries. An important battery type is the so-called dry-cell battery that was first developed by Carl Gassner in 1886 on the basis of a wet zinc-carbon battery that had been developed twenty years earlier by Georges Leclanché. Fig. 7.8 shows a schematic representation of the structure of such a battery, now also known as an 'alkaline battery'. The outer zinc casing contains a slurry of electrolyte contained in a volume bounded by a porous cardboard layer. This electrolyte is composed of two layers: the first is of a paste of ammonium chloride that is positioned next to the zinc anode casing; and the second is a paste of ammonium chloride and manganese dioxide, the latter acting as a cathode. The second layer surrounds a carbon rod set in the centre of the structure.

The anode reaction is:

$$Zn + 2\,Cl^- \rightarrow ZnCl_2 + 2\,e^-$$

and the cathode reaction is:

$$2\,MnO_2 + 2\,NH_4Cl + H_2O + 2\,e^- \rightarrow Mn_2O_3 + 2\,NH_4OH + 2\,Cl^-$$

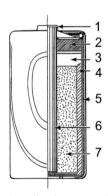

Fig. 7.8 Schematic diagram of a dry cell. 1, brass cap; 2, plastic seal; 3, expansion space; 4, porous cardboard; 5, zinc can; 6, carbon rod; 7, chemical mixture. *(https://en.wikipedia.org/wiki/Dry_cell.)*

so that the overall reaction is:

$$Zn + 2\,MnO_2 + 2\,NH_4Cl + H_2O \rightarrow ZnCl_2 + Mn_2O_3 + 2\,NH_4OH.$$

It is also possible to substitute zinc chloride for the ammonium chloride as the electrolyte when the overall reaction becomes:

$$Zn + 2\,MnO_2 + H_2O \rightarrow Mn_2O_3 + Zn(OH)_2$$

The cell voltage is obtained is 1.5 V but this drops under load and also as a result of the depletion of the reactants. The great advantage of this structure is that it can operate in any orientation without spilling and this allows it to be used in mobile devices such as torches. It requires no maintenance but cannot be recharged.

Rechargeable Ni-Cd and Ni-MH Batteries. The nickel-cadmium (Ni-Cd) and nickel-metal hydride (Ni-MH) batteries are both rechargeable and have voltages that are similar to those provided by the alkaline battery discussed in the previous section: 1.2 V for Ni-Cd and 1.25 V for Ni-MH. As a consequence, both have been commonly used as rechargeable substitutes for the dry-cell batteries described in the previous section. The Ni-Cd battery has been gradually developed over a long period, starting from its original creation in 1899 by the Swede, Waldemar Jungner. By 2000, some 1.5 billion batteries were sold globally per year. The currently available type for portable use is in the form of a Swiss roll with several layers of positive and negative material held in a cylindrical container. Larger variants are also constructed in the form of ventilated cells, these being used for standby power or in electric vehicles (Chapter 6). As cadmium is considered to be a toxic material, its use is now banned in Europe, except for use in a limited number of applications such as medical devices. Furthermore, European regulations require that any producer of these devices is responsible for recycling the cadmium.

The Ni-Cd cell, when producing a current, operates with the following electrode reactions:

$$Cd + 2OH^- \rightarrow Cd(OH)_2 + 2e^-$$

and

$$2NiO(OH) + 2H_2O + 2e^- \rightarrow 2Ni(OH)_2 + 2OH^-$$

The allover reaction is thus:

$$2NiO(OH) + Cd + 2H_2O \rightarrow 2Ni(OH)_2 + Cd(OH)_2$$

During charging, the reaction occurs in the opposite direction. Compared with the dry-cell carbon-zinc batteries, the voltage of the Ni-Cd battery declines very little with the extent of discharge and it can also tolerate very high discharge rates, making it preferable to lead-acid batteries. It is also much lighter than the lead-acid batteries and so is now used in situations where weight is an important factor, for example in aircraft. Because both Ni and Cd are expensive, however, the higher cost of these batteries is also an important factor. As a result, the Ni-Cd battery is currently losing out in most applications to two other types of batteries: either the nickel-metal-hydride (NiMH) that will now be discussed, or the lithium-ion batteries, discussed in the next section.

The Ni-MH battery is a much more recent arrival than the Ni-Cd battery, having been gradually developed over the period since its discovery in 1967. The electrochemical half-cell reactions occurring in the discharging process are:

$$OH^- + MH \rightarrow H_2O + M + e^-$$

and

$$NiO(OH) + H_2O + e^- \rightarrow Ni(OH)_2 + OH^-$$

(Both these reactions occur in the reverse direction during the charging process.) The allover reaction is thus:

$$NiO(OH) + MH \rightarrow Ni(OH)_2 + M$$

Hence, the inter-conversion of $NiO(OH)$ and $Ni(OH)_2$ by reaction with water occurs in both the Ni-Cd and the Ni-MH batteries. However, with the Ni-MH combination, hydrogen is produced and this is absorbed by the Ni-M component of the positive electrode, where M is a rare-earth metal, most commonly Lanthanum (La). It is well established that the alloy $LaNi_5$ readily absorbs hydrogen reversibly to form a stable hydride.[c] Hence, hydrogen gas is not formed within the cell as long as the hydride can be formed. As pure lanthanum is expensive due to the high cost of isolating its oxide from mixtures of the rare earth oxides in which it occurs, so-called mischmetal is often used, this being directly derived from the rare-earth ores without separation and typically having the composition of 55% cerium (Ce), 25% La and 15%–18% neodymium (Nd) in addition to about 5% iron.

---

[c] For a detailed description of the LaNi5 system, see a paper by J. Reilly (Brookhaven National Laboratory) available for download at https://www.osti.gov/servlets/purl/6084207.

(The variability of the composition of the mischmetal explains the rather variable ratios that are listed on various related web sites for the components of commercial Ni-M-H batteries.)

It is important not to overcharge Ni-M-H batteries and hence smart chargers have been developed to prevent the occurrence of such damage. As one of the consequences of overcharging is the evolution of hydrogen, with the consequent possible rupture of the casing, modern cells often contain a catalyst to cause the recombination of this hydrogen with oxygen to form water. Another problem that can occur in a battery pack containing a number of separate cells in series is that if the complete discharge occurs, one or more of these cells may suffer from polarity reversal, this resulting in irreversible damage of the combined arrangement.

One of the great advantages of Ni-MH batteries is that they are much lighter than the Ni-Cd batteries discussed above and another is that they have none of the latter's environmental problems. They are very useful for applications in which high currents are drawn as they have low internal resistance and therefore are less prone to overheat. The Ni-MH battery was the predominant choice for use in early electric vehicles but they have now been largely superseded by lithium batteries. However, patent problems have rather limited their use. It appears that their sale is controlled by Cobasys, a subsidiary of Chevron, which only provides large orders; one consequence of this has been that General Motors has shut down production of its EV1, citing problems with battery availability.

Li-Ion Batteries. The idea of a lithium-based battery system (see Box 7.4) was first investigated in the early 1970s by Stanley Whittingham, then at Stanford University, who showed that it was possible to store Li^+ ions in the layers of a disulphide. He was then hired by Exxon where he showed that it was possible to make a battery consisting of titanium (IV) sulphide and lithium metal as the electrodes. However, this battery was found not to be practicable due to the insurmountable problems such as the need to ensure that the Li did not encounter water and thus release hydrogen gas. Further work discussed in Box 7.4 led to the Li-ion battery as we now know it and, finally, to the award of the 2019 Nobel Prize in Chemistry to Stanley Whittingham (now at Binghampton University), John Goodenough (now at the University of Texas) and Akira Yoshino (now at Meija University, Nagoya and Asahi Kasei Corporation). The citation for the award said of the lithium-ion battery: 'This lightweight, rechargeable and powerful battery is now used in everything from mobile phones to laptops and electric vehicles. It can also store significant amounts of energy from solar and wind power, making possible a fossil-free society.'

BOX 7.4 The 2019 Nobel Prize for Chemistry

The Nobel Prize in Chemistry for 2019 was awarded to John B. Goodenough, M. Stanley Whittingham and Akira Yoshino 'for the development of lithium-ion batteries.' (https://www.nobelprize.org/prizes/chemistry/2019/press-release/).

The work leading to the award to this trio of scientists commenced when Whittingham, who had just completed his postdoc at Stanford University on aspects of intercalation started to work at Exxon where his work involved a search for materials that would intercalate lithium ions for possible battery applications. He found that titanium disulphide, a material with a layered structure, could reversibly accommodate the Li^+ ions. This material is electrically conductive and was found not to interact with other materials such as electrolyte molecules, thus making it ideal as a cathode (in the discharge mode) that would take up Li ions during the discharge process. Exxon funded a major project on the work that led to the development of a battery consisting of a lithium-metal anode, lithium perchlorate in dioxolane as electrolyte, and a titanium disulphide cathode. This battery provided a voltage of 2.5 V and stored about ten times as much energy as a lead-acid battery or five times as much as a Ni-Cd battery. However, it was found that this battery design had serious problems as the lithium deposited spikey formations ('dendrites') that intruded into the electrolyte gap, ultimately causing short circuits that resulted in the combustion of the electrolyte material and even the Li itself.*

Goodenough, a physicist who had previously worked in the Lincoln Laboratory at MIT and is now still working at the University of Texas in Austin, was appointed in 1976 as head of the Inorganic Chemistry Laboratory at Oxford University. He started to work on cathode materials in an effort to replace Whittingham's titanium disulphide. In 1980, he recognised that it might be possible to replace the sulphur species with oxygen. Koichi Mizushima (now of Toshiba Corporation, who was working with him at the time) started to screen a range of metal oxides and found that there existed a series of ternary layer compounds ($LiMO_2$), containing metals such as V, Cr, Fe, Co and Ni, which incorporated Li ions reversibly. He found that $LiCoO_2$ was the most effective of these compounds and that the structure was maintained as the Li ions moved in and out of the structure during the charging and discharging steps. The potential of a battery constructed with a $LiCoO_2$ electrode was significantly increased over that of the Whittingham battery, reaching 4 V. This battery still used a Li metal electrode.

The final step in achieving a usable design was achieved by Yoshino (now at Meijo University, Nagoya) who at that stage worked with Asahi Kasei in Japan. The problem with Goodenough's battery was still that the Li electrode was susceptible to combustion. Yoshino focused on a so-called 'rocking-chair principle' in which the lithium is not converted to metallic species at either electrode but is incorporated into the electrode structures, one of these being

BOX 7.4 The 2019 Nobel Prize for Chemistry—cont'd

the $LiCoO_2$ discussed above. He found that a petroleum-derived coke was an ideal component for the second electrode. Following a number of years of development work by Asahi Kesei, in collaboration with Sony, that was aimed at getting the electrodes into the correct form with adequate strength, the first commercial lithium-ion battery was launched by Sony in 1991.**

* See the section on Lithium Metal Batteries below.

** The fascinating history of the development of the Li-ion battery is covered by a number of useful references, including the Nobel Prize press release referred to above and an article by Katrina Krämer in Chemistry World entitled 'The Lithium Pioneers' (Chemistry World, November 2019, 24–30.)

The Li-ion battery as we now know it has a number of different variants. However, the essential feature of all such batteries that derive from the work of Whittingham, Goodenough and Yoshino is that Li^+ ions are transported to and from the electrodes by reducing the cobalt species in a $Li_{1-x}CoO_2$ material from Co^{4+} to Co^{3+} during the discharge process and re-oxidising the Co^{3+} to CO^{4+} during the charging process. The reaction occurring at the positive electrode (cathode) during the discharging reaction can be depicted as follows:

$$CoO_2 + Li^+ + e^- \rightarrow LiCoO_2$$

At the negative electrode, Li metal, intercalated in a graphite matrix, is oxidised in the reaction:

$$LiC_6 \rightarrow C_6 + Li^+ + e^-$$

The allover discharging reaction is thus:

$$LiC_6 + CoO_2 \rightarrow C_6 + LiCoO_2$$

These reactions are reversible and occur in the opposite direction in the charging process. Water cannot be used as a solvent for the electrolyte by which the Li + ions are transported since hydrolysis of the water to hydrogen and oxygen would then be preferred to the liberation of lithium metal at the CoO_2 electrode. Hence, a variety of different organic solvents have been used, these including ethylene carbonate, dimethyl carbonate and diethyl carbonate. Recently developed variants use solid ceramic ion conductors such as perovskites for the transport of the Li^+ ions. The Li^+ ions are generally included in lithium salts such as $LiPF_6$, $LiBF_4$ or $LiClO_4$, all of which are soluble in the organic solvents mentioned above.

Charging of the Li-ion battery must be carefully controlled by limiting the peak voltage in order to avoid irreversible damage. For example, overcharging at relatively low voltages causes supersaturation of the lithium cobalt oxide electrode material, thus resulting in the formation of lithium oxide by the irreversible reaction:

$$Li^+ + e^- + LiCoO_2 \rightarrow Li_2O + CoO$$

Overcharging at higher voltages of up to 5.2 V results in the production of Co(IV) oxide by the following reaction:

$$LiCoO_2 \rightarrow Li^+ + CoO_2 + e^-$$

and this again occurs irreversibly.

Another limitation of the Li-ion system is that the Co electrode reaction described above is only reversible for $x < 0.5$ and this limits the depth of discharge allowable. It is therefore critical in the use of a Li-ion battery that charging and discharging are very carefully controlled and it is therefore necessary to use specially designed charging systems and also to monitor and control the discharge process. Nevertheless, even though Li batteries are more expensive than the Ni-Cd batteries discussed in the previous section, they have the advantage that they operate over a wider temperature range and give higher energy densities. They are also significantly lighter, an important advantage in applications such as their use to power electric vehicles as discussed in Chapter 6.

Lithium Metal Batteries. Also depending on the Li chemistry discussed above and also based on the work of Whittingham in Exxon is the class of cell referred to as a lithium metal battery. These batteries are irreversible and therefore disposable, consuming the lithium metal to form Li ions. They are able to provide high power and are commonly preferred for use in small portable electronic devices such as watches and digital cameras. They generally have long lives and, although more expensive, are therefore often used in place of the more common alkaline cells described in an earlier section. Because of the danger that the lithium metal in the cell can react with water vapour to produce explosive hydrogen mixtures if the cell is damaged, restrictions are in place for the transport of these lithium metal batteries. They are also susceptible to the growth of dendrites in the non-aqueous electrolyte between the Li anode and the cathode and this can cause short-circuiting and failure of the cells, a problem that can be avoided with the correct choice of electrolyte. The interested reader should search the web for further information on lithium metal batteries and how they are distinguished from lithium-ion batteries.

Flow batteries

In the battery types described above, the reactants providing the power are components of the constituent cells. In contrast, flow batteries depend on the continuous supply of reactants from outside the assembly and the simultaneous continuous collection of the products as shown schematically in Fig. 7.9.

The cell works in exactly the same way as an electrochemical cell as described in a previous section apart from the continuous replacement of reactant. For a redox cell, the reactants are dissolved in the electrolyte on one side of the cell and the products of the electrochemical reaction are collected in the electrolyte on the other side of a separating proton exchange membrane. Many different combinations of reactants have been examined for use and these include the Mn^{6+}/Mn^{7+} and V^{3+}/V^{5+} redox couples. For the vanadium cell, four different oxidation states occur, the electrode reactions occurring during discharge at the positive electrode being:

$$VO_2^+ + 2H^+ + e^- \rightarrow VO^{2+} + H_2O$$

and the simultaneous reaction at the negative electrode being:

$$V^{2+} \rightarrow V^{3+} + e^- \text{ (negative electrode)}$$

The allover discharge reaction is therefore:

$$VO^{2+} + 2H^+ + V^{2+} \rightarrow VO^{2+} + H_2O + V^{3+}$$

All the reactions shown are reversible and the recharging reaction, therefore, goes in the opposite direction. During both discharging and

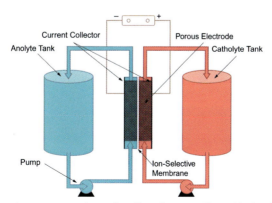

Fig. 7.9 Schematic representation of a flow battery. *(https://upload.wikimedia.org/wikipedia/commons/5/5b/Redox_Flow_Battery.jpg.)*

charging, a proton is transferred across the membrane and an electron is transferred between the electrodes. A typical open-circuit voltage at room temperature is 1.41 V. Typical electrodes consist of a felt or cloth made of either carbon or graphite.

The advantage of such a set-up is that the extents of the reactions occurring are determined not by the contents of the cell but by the amounts of reactants and products that can be transmitted through the cell assembly and so they depend only on the volume of the storage tanks. Such systems are used for stationary applications such as load balancing for an electrical grid, when excess production of power can be transferred by the electrochemical process to chemicals, these being stored until power is required when it is generated by operating the reverse process. The advantage of such a system is that there is none of the gradual loss of electrical energy due to leakage that there would be with a battery storage system. However, all-over energy losses do still occur due to the need to supply energy for the pumps in the so-called 'analyte' and 'catholyte' storage systems.

Fuel cells

A fuel cell has a very similar function to a flow battery but it is generally used to convert fuel directly and irreversibly to electrical energy. Fuel cells are most commonly used for the oxidation of hydrogen (forming water) although they can also be used for the conversion of a number of other fuels. The structure of a fuel cell for use with hydrogen is shown schematically in Fig. 7.10.

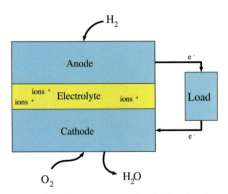

Fig. 7.10 A block diagram for the structure of a fuel cell using hydrogen as a fuel. *(https://upload.wikimedia.org/wikipedia/en/1/1b/Fuel_Cell_Block_Diagram.svg.)*

Hydrogen and oxygen (or air) are fed separately to the anode and cathode compartments respectively. Hydrogen dissociates at the surface of the anode and loses two electrons, forming H^+ ions:

$$H_2 \rightarrow 2\,H^+ + 2\,e^-$$

The hydrogen ions then transfer through the electrolyte material, the nature of which depends on the type of fuel cell in use (see below), and they then combine with oxygen at the cathode, producing water:

$$2\,H^+ + 0.5\,O_2 \rightarrow H_2O$$

The allover reaction is thus:

$$2\,H_2 + O_2 \rightarrow 2\,H_2O$$

As the half-cell potential for the reaction $O_2 + 4\,H^+ \rightarrow 2\,H_2O$ is 1.229 V and that for the hydrogen ionisation reaction is 0.0 V, a typical single hydrogen fuel cell produces a voltage of between 0.6 and 0.7 V, this dropping with increasing load. A series of such cells can be built into a stack with the result that significantly higher voltages can be achieved, delivering powers up to several MW. Heat is also produced and so fuel cells can be used very effectively for combined heat and power applications.

The electrolyte in the fuel cell allows the transport of the cations, in most cases H^+, from the anode to the cathode. For the simplest type of fuel cell, first described by Sir William Grove in 1838, the electrolyte was simply a solution of copper sulphate and dilute sulphuric acid while the electrodes were sheets of iron and copper. This design was similar to what is now known as the phosphoric acid fuel cell in which the phosphoric acid replaces the sulphuric acid used by Grove. The various types of fuel cells generally available, as well as their uses, are summarised in Table 7.3. As well as in military and space applications, for which much of the background development work was carried out by NASA and other space agencies, the principal uses of all these fuel cell types are in backup power and aspects of electricity distribution. The limited use of transportation was discussed in Chapter 6.

Polymer electrolyte membrane fuel cell (PEM). The construction of all the different types of fuel cell assemblies are similar and they differ mostly in the nature of the electrolyte that transfers ions between the electrodes. The structure of the polymer electrolyte membrane (PEM) fuel cell is shown in Fig. 7.11.

Hydrogen ions are formed at the anode, which is composed of platinum, and these are then transported through the polymer electrolyte membrane to

Table 7.3 Fuel cell types.

Fuel cell type/abbreviation	Electrolyte	Operating temperature/°C	Electrical efficiency/%	Applications
Alkaline fuel cell (AFC)	Aqueous KOH or alkaline polymer membrane	<100	60%	Military Space Backup power Transportation
Polymer electrolyte membrane fuel cell (PEM)	Perfluorosulphonic acid	<120	60%	Backup power Portable power Distributed generation Transportation Speciality vehicles
Phosphoric acid fuel cell (PAFC)	Phosphoric acid in a porous matrix or in a polymer membrane	150–200	40%	Distributed generation
Molten carbonate fuel cell (MCFC)	Molten lithium, sodium and/or potassium carbonates in a porous matrix	600–700	50%	Electric utility Distributed generation
Solid oxide fuel cell (SOFC)	Yttria stabilised zirconia	500–1000	60%	Auxiliary power Electric utility Distributed generation

Batteries, fuel cells and electrolysis   189

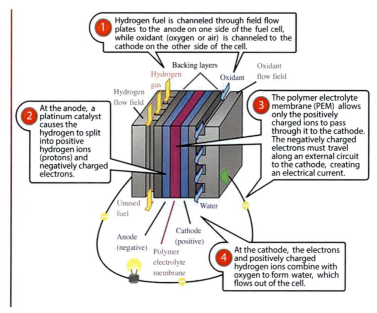

Fig. 7.11 A cut-away diagram of a polymer electrode membrane fuel cell. *(https://upload.wikimedia.org/wikipedia/commons/0/0d/PEM_fuelcell.svg)*

the cathode where they combine with $O^=$ ions to form water, this then flowing out of the cell. As the platinum electrode material can be irreversibly poisoned by the adsorption of impurities in the feed gas, particularly of any CO remaining from the steam reforming process used to create it, the feed must contain very low concentrations of such impurities. However, this restriction becomes significantly less important as the temperature of operation of the fuel cell is increased. For example, the molten carbonate fuel cell, operating at about 650°C, is much less susceptible to CO poisoning.

A relatively recent application of the PEM is the Direct Methanol Fuel Cell (DMFC) in which an aqueous solution of the methanol fuel is fed to a cathode containing platinum and ruthenium which then reforms the methanol–water mixture directly to give hydrogen and carbon dioxide. The allover anode reaction can therefore be expressed as follows:

$$CH_3OH + H_2O \rightarrow 6\,H^+ + 6\,e^- + CO_2.$$

The DMFC is most commonly used currently in place of more conventional batteries in applications such as fork-lift trucks used in warehouses but it has potential for use in automobiles if the structure exists for the distribution of the methanol fuel. Another possible fuel for such purposes is formic acid which can be decomposed directly in situ to give hydrogen and CO_2. As both methanol and formic acid can be obtained from non-fossil fuels, their use would not contribute to the emission of greenhouse gases. Such an approach is discussed further later in this chapter.

Molten carbonate fuel cell (MCFC). In the molten carbonate fuel cell (see Fig. 7.12), the electrolyte is a mixture of molten carbonates, Li_2CO_3, K_2CO_3 and/or Na_2CO_3, suspended in a ceramic matrix. Carbonate ions are formed at the cathode and pass through the carbonate mixture of the matrix to the anode where they react with syngas (carbon monoxide and hydrogen) to give carbon dioxide and water.

The reactions occurring are as follows:

Anode reaction:

$$H_2 + CO_3^= \rightarrow H_2O + CO_2 + 2e^-$$

Cathode reaction:

$$1/2 O_2 + CO_2 + 2e^- \rightarrow CO_3^=$$

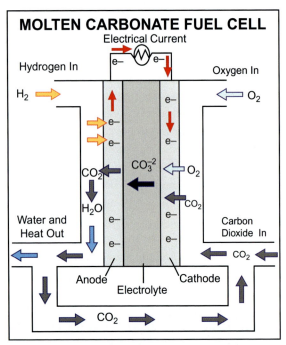

Fig. 7.12 A molten carbonate fuel cell. *(United States Department of Energy.)*

The cell reaction is thus:

$$H_2 + 1/2 O_2 \rightarrow H_2O$$

The high temperature of operation, between 600°C and 700°C, increases the rates of all the reactions involved and this means that it is not necessary to use noble metals such as platinum for the electrodes. The anode thus generally comprises of an alloy of nickel-containing chromium or aluminium and the cathode is either lithium metatitanate or NiO intercalated with lithium species. Because these electrodes are much less prone to poisoning by CO, a variety of different fuels can be used, including fuel gas (obtained from coal), methane or natural gas. The methane can be reformed to syngas just prior to admission of the fuel to the system (the reaction being termed 'Internal Reforming') in which case the fuel reacting the cathode is approximately a 3:1 mixture of H_2 and CO:

$$CH_4 + H_2O \rightarrow CO + 3H_2$$

The oxidant fed to the cathode compartment is either O_2 or CO_2; in the latter case, the CO_2 formed at the anode in the anode half-cell reaction described above is circulated back to the cathode. Some problems have been encountered due to poisoning and corrosion as a result of passage of the alkali metal ions present in the cell to other parts of the system but these difficulties have largely been overcome. The use of MCFC's is predominantly in large systems such as fuel-cell powered electricity generation stations and combined heat and power plants.

Solid oxide fuel cell (SOFC). The construction of the solid oxide fuel cell is shown schematically in Fig. 7.13. In this, the oxygen fed to the

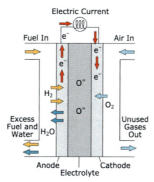

Fig. 7.13 A solid oxide fuel cell. *(https://upload.wikimedia.org/wikipedia/commons/4/42/Solid_oxide_fuel_cell.svg.)*

192   Sustainable energy

cathode compartment is transmitted as $O^=$ ions through a ceramic membrane to the anode where these ions oxidise the fuel directly.

The fuel is generally hydrogen but it could also be methane, propane or butane or even gasoline, diesel or jet fuel. The membrane material is a dense ceramic that conducts oxygen ions without having any electronic conduction ability, normally yttrium-stabilised zirconia. The anode is commonly a cermet comprising nickel mixed with the electrolyte material. It performs the function of not only providing the H^+ ions reacting with the $O^=$ ions diffusing through the electrolyte but also reforming the hydrocarbon fuels that are frequently fed instead of pure hydrogen. The cathode material is currently lanthanum strontium manganite, compatible with the yttria-doped zirconia electrolyte, its function being to convert the oxygen feed into $O^=$ ions. The operation temperature is generally high, between 500°C and 1000°C, to facilitate the transport of the $O^=$ ions through the solid electrolyte membrane layer. The main current use of SOFCs is in applications such as stationary power generation, the systems having outputs up to 2 MW. There is currently much research and development work related to the application of SOFCs as well as to the other types of the fuel cell. Some of the most recent work is discussed in Chapter 8.

Electrolysis

The process of electrolysis is most commonly encountered in relation to the production of hydrogen from water. In this case, the reaction is the reverse of that encountered in the operation of the alkaline and polymer electrolyte membrane fuel cells discussed in the previous section. However, electrolysis has many other applications, some of which will also be discussed briefly below.

The electrolysis of water has been known for more than 200 years, much of the early work having resulted from the invention of the Volta pile described above. In common with all other important applications of electrolysis, the electrolysis of water is a reaction that is not allowed thermodynamically to occur without the application of an external voltage. Currently, the cost of hydrogen produced using electrolysis is higher in most developed economies than that of the hydrogen from fossil-fuel-based steam reforming plants of the type discussed in Chapter 2. However, with the steady progress that is currently occurring towards the introduction of renewable energies (Chapter 3) and the consequent reduction of the cost of electricity, it is clear that electrolysis will soon become a competitive and more acceptable route

Batteries, fuel cells and electrolysis   193

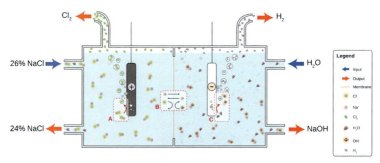

Fig. 7.14 The membrane chloralkali process. (https://upload.wikimedia.org/wikipedia/commons/7/7a/Chloralkali_membrane.svg.)

for the production of hydrogen for a wide range of different applications, including for the operation of fuel cell systems. Some of the potential uses of 'green' hydrogen will be discussed in more detail in Chapter 8. For the purposes of the current chapter, it is sufficient to say that the types of electrolyses currently being used are very closely related to the fuel cells described above. In many ways, fuel cells could have been described under the heading 'Reversible Batteries' as they can all occur reversibly when suitable potentials are applied. Electrolysis is a well-established approach to obtaining many of our currently important industrial products. For example, the production of chlorine and sodium hydroxide from brine can be carried out using a number of different technologies, one of the most modern of which is the membrane cell electrolysis process shown schematically in Fig. 7.14.

In this, a solution of sodium chloride is fed to the anode compartment and some of the sodium ions are transmitted through the membrane to the cathode compartment, the sodium ion concentration of the brine solution leaving the anode compartment having been reduced. Chlorine gas is formed simultaneously and is emitted from this compartment. The sodium ions entering the cathode compartment react with hydroxyl ions formed at the cathode and hydrogen is liberated simultaneously. The cathode reaction is:

$$2\,H^+_{(aq)} + 2\,e^- \rightarrow H_{2\,(g)}$$

and the anode reaction is:

$$2\,Cl^-_{(aq)} \rightarrow Cl_{2(g)} + 2\,e^-$$

and so the allover reaction is:

$$2\,NaCl + H_2O \rightarrow Cl_2 + H_2 + 2\,NaOH$$

The process can also be carried out with KCl to produce KOH.

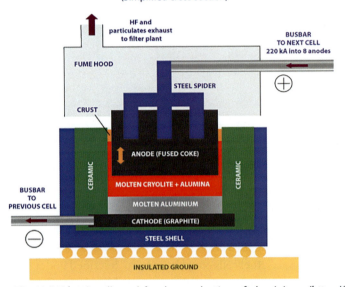

Fig. 7.15 The Hall-Hérault cell used for the production of aluminium. *(https://upload.wikimedia.org/wikipedia/commons/2/24/Hall-heroult-kk-2008-12-31.png)*

Other electrolysis processes of importance include the production of aluminium from alumina by the Hall-Hérault process in which a molten mixture of alumina with cryolite (Na_3AlF_6) and calcium fluoride is electrolysed at temperatures between 950°C and 980°C; see Fig. 7.15.

The system consumes the carbon of the anode. The cathode reaction is:

$$Al^{3+} + 3e^- \rightarrow Al$$

and the anode reaction is:

$$O^{2-} + C \rightarrow CO + 2e^-$$

so that the overall reaction is:

$$Al_2O_3 + 3C \rightarrow 2\,Al + 3\,CO$$

In practice, CO_2 is also formed at the anode, the allover reaction then being:

$$2\,Al_2O_3 + 3C \rightarrow 4\,Al + 3\,CO_2.$$

The production of aluminium consumes large amounts of electrical energy and so contributes very significantly to the production of greenhouse gas unless renewable electrical energy is used. Many other metals are also produced from their oxides or other related raw materials by electrolysis processes. These include magnesium, zinc, lead, chromium, manganese and titanium. A useful summary of some of these processes is to be found at https://www.osti.gov/servlets/purl/6063735.

CHAPTER 8

The way forward: Net Zero

Introduction

Chapter 1 outlined the problems associated with greenhouse gas emissions and the resultant global warming which is currently occurring and went on to discuss attempts that are being made to combat this warming. The Paris Accord of 2015, to which the majority of the countries in the world have subscribed, was drawn up under the auspices of the United Nations Convention on Climate Change (UNFCCC). This accord has as its aim the achievement of a limit to the increase in global temperature since pre-industrial levels of not more than 2.0°C; a secondary aim of the accord is that every effort should be made to achieve a lower temperature increase of only 1.5°C. The term 'Net-Zero' has recently been introduced, this corresponding roughly to the aim towards the 1.5°C limit of the Paris Accord. A very recent report[a] from the International Energy Agency, entitled 'Net Zero by 2050—A Roadmap for the Global Energy Sector', supplies pointers as to what must be achieved if the global community is to achieve the Net-Zero target by 2050 and some of its contents are very relevant to the material of this book. Fig. 8.1, taken from that report, shows the various hurdles that must be surmounted in the intervening period. The report stresses that each participating government will need to set its own targets since each country will have differing needs and opportunities and that much will depend on the commitment of all sectors of society in each region of the global community to achieve these targets. It is already quite clear that many countries are already falling far behind the targets that they had set themselves in their action plans drawn up in response to the Paris Accord and that significant changes need to be made if we are even to achieve the higher 2.0°C target.

Some significant pointers from the boxes of Fig. 8.1 are summarised below.
- From today, no new oil or gas fields or coal plants should be approved for development.
- From 2025, no fossil fuel boilers should be sold.

---

[a] https://www.iea.org/reports/net-zero-by-2050.

Sustainable Energy
https://doi.org/10.1016/B978-0-12-823375-7.00005-6

Copyright © 2022 Elsevier B.V.
All rights reserved.

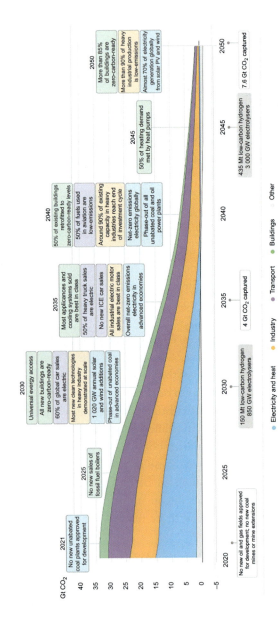

Fig. 8.1 Net zero by 2050. (From Net Zero by 2050—A Roadmap for the Global Energy Sector. https://www.iea.org/reports/net-zero-by-2050)

- By 2030, all new buildings should be zero-carbon ready; 60% of car sales should be electric; there should be substantial progress in developing solar and wind electricity generation; and the use of coal without CO_2 capture and storage should be terminated.
- By 2035, 50% of heavy-duty trucks should be electric; there should be no further sales of vehicles with internal combustion engines; and all advanced economies should have only renewable electricity.
- By 2040, 50% of existing buildings should be retrofitted to zero-carbon levels; 50% of fuels for aviation should be low emission; 90% of the heavy industrial equipment now in use should have been replaced; all electricity globally should be renewable; and all unabated coal- and oil-based power plants should have closed down.
- By 2045, 50% of heating should be supplied by heat pumps.
- By 2050, more than 85% of buildings should have reached zero-carbon levels; more than 90% of the heavy industry should use low-emission technology; and almost 70% of electricity globally should be renewable.

Two other important topics featured in the IEA report are shown at the base of Fig. 8.1: low-carbon electrolysers and CO_2 capture. These two topics will now be discussed further in the light of some of the material of Chapters 1–7. This chapter then concludes with a brief mention of several other topics of relevance to the reduction of greenhouse gas emissions.

Hydrogen production using renewable energy

It is now generally accepted that the most desirable approach to enabling the reduction of greenhouse emissions is the generation of 'green' hydrogen by electrolysis of water using renewable electricity. However, it is also possible to produce 'blue' hydrogen by steam reforming of raw materials such as natural gas as described in Chapter 4 if successful methods can be established for the complete capture and successful storage of all the CO_2 formed during the reforming process. Currently, as shown in Box 8.1, the costs of producing hydrogen by conventional electrolysis methods using alkaline cells or polymer electrode cells are currently much higher than those for the production by the steam reforming of methane. However, for some purposes, electrolysis using these methods is already the preferred route when the price of carbon emissions is factored in. As will be discussed further in some detail in a later section, there is also some hope that the use of high-temperature hydrolysis using solids oxide cells of the type currently under development may become commercialised very soon.

200   Sustainable energy

BOX 8.1 The cost of producing hydrogen by electrolysis

A recent report (2020) prepared for the European Parliament entitled 'The Potential of Hydrogen for Decarbonising Steel Production' gives some of the costs of producing hydrogen.[b] The cost of 'grey' hydrogen produced by steam reforming varies quite significantly and depends on the cost of the natural gas used. Table 4.1 gives the current prices for hydrogen production using the different methods described in Chapter 4, these ranging from about €1.3–2.0/kg. In the electrolysis process, about 70%–80% of the electricity used is required to produce 'green' carbon-dioxide-free hydrogen and this requires 50–55 kWh (kWh) of electricity per kg. The cost of green hydrogen is thus currently in the range of €3.6–5.3/kg, depending on the exact cost of electricity in the country in question. The price of electricity has fallen by 60% over the last decade and it is predicted that it will continue to fall as the technology for electrolysis is improved and the costs of renewable energy decrease. The report, therefore, suggests that the price of green hydrogen production by electrolysis in Europe could fall as low as €1.8/kg by 2030, although the basic costs will vary somewhat in non-European countries. For example, in the US, where the cost of natural gas produced by fracking is significantly lower[c], electrolysis may be less competitive even when there is a reduction in the cost of renewable energy. In contrast, as will be discussed further below in relation to the costs of steel production using hydrogen, it may soon be commercially attractive for Australia to produce the necessary hydrogen by electrolysis.

[b]https://www.europarl.europa.eu/RegData/etudes/BRIE/2020/641552/EPRS_BRI(2020)641552_EN.pdf.
[c]A recent report from the Fuel Cell and Hydrogen Energy Association (FCHEA). https://www.fchea.org/us-hydrogen-study discusses the importance to the US of hydrogen produced from fracked natural gas by steam reforming, the carbon dioxide produced being captured and stored. The report, compiled by McKinsey, includes details of storage sites as well as showing the infrastructure already existing for the transportation of hydrogen by pipeline. Although a similar infrastructure exists in parts of Europe, this is not the case globally.

Chapter 3 discussed the many different routes to produce renewable energy that have been either fully developed or are currently being investigated. It is clear that in order to reach Net-Zero by 2050, some of these renewable resources must be developed even more rapidly than hitherto and that each jurisdiction must do everything possible to develop these resources in such a way that it will be possible to phase out the use of fossil fuels in the vast majority of current applications. Oil will still most probably have to be used as a source of chemicals for use, for example, in the production of polymers and pharmaceuticals. However, fossil-fuel-based petroleum and diesel will be things of the past, superseded in some instances by fuels derived from biomass. Although nuclear energy is

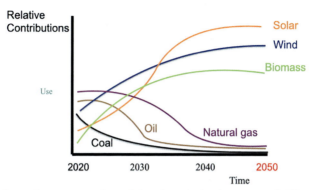

Fig. 8.2 Schematic representation of the changes in the usage of different energy sources that will be required by the year 2050 to achieve Net Zero.

currently considered to be a contributor to the world's renewable energy resources, the problem of the disposal of nuclear waste is insurmountable and it is likely that the number of countries banning the use of nuclear power will increase. It is likely therefore that the majority of the world's economies will rely on the production of their renewable electrical energy using either wind or solar generation. Some countries have significant hydro-electric power and one or two have geothermal resources so that these methods will in these cases contribute to the energy mix. Tidal and ocean current generation systems, although available on some seaboards, are likely only to have very limited application due to the difficulties of operating in what are often extreme conditions. The use of coal for the production of electricity in countries such as Australia, China and India, where large reserves of coal are available, will only be acceptable if reliable methods are found to trap and store the CO_2 produced and hence much work remains to be done on developing and testing appropriate methods for carbon capture and storage.

Fig. 8.2 shows schematically the decreasing use of coal, oil and natural gas that might be expected over the next 30 years if Net Zero is to be achieved and it also sketches the expected growth in the same period of the use of wind, solar and biomass for the production of energy. As the use of coal will only be acceptable in situations in which its use is unavoidable, it is expected that Australia, China and India will before 2050 have installed sufficient wind and solar power resources to enable them to minimise their use of coal.[d] Oil will no longer be used for traction and natural gas will only be

---

[d] A very useful source of information on the continuous progress of the installation of solar and other green power sources is provided by a newsletter updated daily: daily.newsletter@pv-magazine.com. This newsletter also gives authoritative information on developments in a hydrogen economy.

202 Sustainable energy

used when adequate CO_2 capture facilities exist. The net result will be that many products currently made using coal, oil or natural gas will be produced by new routes while biomass–derived fuels will substitute some of the uses of oil–derived products for traction purposes. Box 8.2 lists some of the national reports on possible developments in renewable energy provision and use. Of

---

BOX 8.2 Some national reports on the use of hydrogen

There exist a number of organisations and networks dedicated to the development of the use of hydrogen and of fuel cell technologies. As a result, a large number of reports have been written on the subject. Some of these are described briefly below. Hydrogen Europe (https://www.hydrogeneurope.eu) is an organisation that incorporates a range of some 260+ European industrial companies and 27 national associations. Partnered with the European Commission, it is involved in an innovation programme entitled 'Clean Hydrogen for Europe'. The European Commission has established a fund of €10 billion with the aim of speeding up 'the transition towards a green, climate neutral and digital Europe'. Together with the Hydrogen Alliance (https://www.ech2a.eu), the Commission will devote part of this funding to achieving the aims given in a report entitled 'A Hydrogen Strategy for a Climate-Neutral Europe' (https://ec.europa.eu/energy/sites/ener/files/hydrogen_strategy.pdf). The priority set forward in this report is the achievement of renewable hydrogen formed by electrolysis using largely wind or solar energy but in the short and medium term, also other forms of low-carbon hydrogen.

In the US, a group comprising major oil companies, car makers, hydrogen producers and fuel cell manufacturers is pushing the US government to follow the lead of Europe and has published a report compiled by McKinsey and Company entitled: 'Road Map to a US Hydrogen Economy' (https://www.fchea.org/us-hydrogen-study). This report emphasises that hydrogen has many advantages as an energy carrier and urges that the US government invest funding for research, development, demonstration and deployment of hydrogen technologies. It stresses that 'directing capital to hydrogen is key to enabling growth in the US'.

Japan, despite having no hydrogen-based industry of its own, has also developed a strategy for hydrogen usage. This includes the goals of demonstrating the storing and transportation of hydrogen from abroad by 2022, introducing full-scale hydrogen generation by around 2030 and achieving fully-fledged domestic use of CO_2-free hydrogen by 2050. Japan aims to have 200,000 fuel-cell vehicles by 2025 and 800,000 by 2030, with 320 refueling stations by 2025 and 900 by 2030 (https://www.meti.go.jp/english/press/2017/pdf/1226_003b.pdf). For additional useful information on the Japanese strategy, see a report prepared by the New Zealand Embassy in Tokyo: https://www.mfat.govt.nz/en/trade/mfat-market-reports/market-reports-asia/japan-strategic-hydrogen-roadmap-30-october-2020/.

Continued

BOX 8.2 Some national reports on the use of hydroge—cont'd

An interesting report compiled by the International Energy Association and the Clingendael International Energy Programme entitled 'Hydrogen in North-Western Europe' (https://www.iea.org/reports/hydrogen-in-north-western-europe) considers how the countries of this region, some of which are relatively far advanced in achieving large supplies of renewable energy through hydroelectric schemes and off-shore wind power, can collaborate and benefit from hydrogen developments in their neighbouring countries. Partners in the programme are Belgium, Denmark, France, Germany, The Netherlands, Norway and the United Kingdom. Part of the programme envisaged will develop the use of carbon dioxide capture in off-shore oil wells in the North Sea and elsewhere.

particular interest in the current context are the many potential applications of 'green hydrogen' to be produced by electrolysis methods that are discussed in some of these reports.

Fuel cells to be used for transportation purposes

The topic of fuel cells for use in automobiles and other vehicles was discussed in Chapter 6 for the situation in which the fuel of choice was hydrogen gas, this being transported and stored under pressure. As discussed in that chapter, a severe limitation of the use of hydrogen in such an application is that there is no existing infrastructure for the supply of hydrogen other than in very specific locations such as California. Despite the hope that was prevalent several decades ago that molecular hydrogen would become widely used in transport, the rapid rise in the availability of electric battery vehicles, tied to parallel major advances in battery technology over the same period, has meant that the widespread use of hydrogen as a fuel is now much less likely, at least in the immediate future. As a result of the hope that hydrogen might be used, there has been a very significant research effort devoted to ways in which hydrogen might be transported and stored so that it could be used in transport applications. One possibility under consideration has been that it could be stored in the form of a metallic hydride or another similar material.[e] If a material could be found that had a sufficiently high storage capacity, its use would be much preferable

---

[e] A following is a useful review on hydrogen storage materials: 'Hydrogen storage: the major technological barrier to the development of hydrogen fuel cell cars', D.K. Ross, Vacuum, 80 (2006) 1084–1089. A preprint of this paper is available at http://usir.salford.ac.uk/id/eprint/16768/.

to the use of sturdy and weighty high-pressure vessels used in the currently available fuel-cell vehicles described in Chapter 6. If such storage materials could be developed, the distribution of hydrogen would also become much simpler.

There remains the possibility that fuels other than hydrogen might be used in fuel-cell vehicles. The most highly advanced of such options is the use of methanol as fuel in a methanol fuel cell, of the type mentioned briefly in Chapter 7, which has already been used in a number of demonstration vehicles. Fig. 8.3 shows the all-over construction of a methanol fuel-cell automobile operating with a PEM fuel cell. Some of the constituent parts are also depicted, including a methanol-reformer used to convert the methanol to hydrogen and CO_2.[f]

The reaction occurring in methanol reforming is:

$$CH_3OH + H_2O \rightarrow 3H_2 + CO_2$$

and this takes place in a separate reactor prior to the entrance to the fuel cell. If the methanol is synthesised from renewable hydrogen using, for example, bio-derived CO_2, the use of this approach can be considered to give 'green' or 'Zero Carbon' power.[g] There is currently great interest in using methanol fuel-cells in other applications such as in freight transport, shipping and stationary applications. The work referred to in footnote 'f' was done in an EU-funded project in which the aim was to provide a catalyst for use in a fuel cell for a ship to be operated on Lake Como.

A methanol-fuelled vehicle has some very significant advantages over both electric vehicles and ones powered by hydrogen fuel cells. Some of the advantages to be gained by using fuel cells fed with methanol or with similar fuel cells fuelled by ethanol or formic acid (see Chapters 5 and 7) are given in Box 8.3.

---

[f] Catalysts for the methanol reforming reaction are generally Cu-containing and there has been significant research devoted to improving their behaviour. A relevant paper from the author's laboratory, 'Methanol reforming for fuel-cell applications: Development of zirconia-containing Cu-Zn-A catalysts' (J.P. Breen and J.R.H. Ross, Catal. Today, 51 (1999) 521–533) describes the preparation and properties of a particularly active and stable catalyst.

[g] As is described below, another route to green methanol uses an SOEC system to produce the syngas used.

BOX 8.3 Advantages of fuel cells powered by methanol

- The range of a car powered by a methanol fuel cell (ca. 1000 km; see Fig. 8.3) is significantly higher than that of most, if not all, battery vehicles and probably also of hydrogen fuel-cell vehicles.
- The refueling time is very short, roughly equivalent to that of a petrol-fuelled vehicle (ca. 3 min), this being much lower than the average recharge time required for battery vehicles or for refueling with high-pressure hydrogen.
- Although there is currently no infrastructure for the provision of methanol, it would require little modification to the current fuel distribution system to allow methanol to be distributed and for suitable pumps to be installed in every refilling station.
- Methanol can be handled more safely than can gaseous hydrogen. It disperses rapidly if spillage occurs, either evaporating or dissolving in water. Although methanol is poisonous if swallowed, it should be recognised that ingestion of petrol is also dangerous.

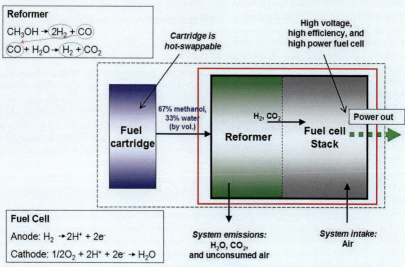

Fig. 8.3 A methanol fuel cell vehicle. (https:www.methanol.orgwp-contentuploads 202004Methanol-Fuel-Cell-Powering-the-Future-webinar-presentation.pdf.)

Solid oxide hydrolysis cells (SOEC's) for hydrogen production and their use for the synthesis of green ammonia and methanol

The production of hydrogen by the electrolysis of water using electrical energy avoids the need for an elaborate hydrogen supply network to provide hydrogen for use in more isolated industrial complexes since existing

206 Sustainable energy

electrical grid connections could be used to transmit electricity to the site at which the hydrogen is required.[h] Such a possibility has focussed attention on the use of 'green' hydrogen in non-transport-related applications that have previously been dominated by the production of 'grey hydrogen' from coal, oil or natural gas (or 'blue hydrogen' produced using carbon capture and storage). Two such applications are the production of 'green ammonia' and 'green methanol'.

Green Ammonia

As discussed in Chapter 4, ammonia is prepared commercially from nitrogen and hydrogen using the Haber-Bosch Process. Chapter 4 also includes a detailed description of the Haldor Topsøe process using hydrogen made from natural gas (Figs. 3.14–3.16). Reports from Haldor Topsøe have recently shown that this process can easily be adapted to use 'green hydrogen' and have also described work on this development. Solid Oxide Electrolysis Cells (SOEC's) for the electrolysis of water to give the needed hydrogen. The SOEC is similar in structure to the Solid Oxide Fuel Cell (SOFC) discussed in Chapter 7. It consists of two permeable solid electrodes separated by a solid oxide membrane that allows transport only of $O^=$ ions formed by dissociation of water from the feed at the negative electrode, this generally being an oxide material containing Ni species:

$$2\,H_2O + 4\,e^- \rightarrow 2H_2 + 2O^=$$

The $O^=$ is then discharged as oxygen gas at the positive electrode, this typically, for higher performance applications, containing mixed conductors such as lanthanum-strontium-ferrite-cobaltite (LSCF) or lanthanum-strontium-cobaltite (LSC):

$$2\,O^= \rightarrow O_2 + 4\,e^-$$

The operation temperature is 600°C to 850°C and, in consequence, the process is much more efficient than the equivalent one occurring in the lower temperature range used for either an alkaline hydrolysis system or a polymer electrolyte membrane hydrolysis system.

---

[h] The grid structures in many countries will still need some modification since the sources of renewable energy such as off-shore wind are often confined to regions well away from the major industries. However, the costs of such modifications are likely to be low in comparison to the provision of additional pipelines. New industries using renewable energy may also become sited closer to the energy sources.

Haldor Topsøe has recently established a demonstration plant for the production of ammonia using their SEOC system. Fig. 8.4 shows a single SOEC assembly of the type used, Fig. 8.5 shows an array of several cells assembled together and Fig. 8.6 shows the complete ammonia synthesis reactor assembly.

Fig. 8.4 A single SEOC cell from Haldor Topsøe. *(Reproduced with the kind permission of Haldor Topsøe.)*

Fig. 8.5 An array of Haldor Topsøe SEOC cells. *(Reproduced with the kind permission of Haldor Topsøe.)*

Fig. 8.6 A complete demonstration plant for ammonia synthesis based on Haldor Topsøe's SEOC units. *(Reproduced with the kind permission of Haldor Topsøe.)*

The possibility of using ammonia as a fuel has recently gained significant attention. In a recent article in Lloyd's Register, Charles Haskell has reported that there are advanced plans for the introduction of ships fuelled by ammonia.[i] The possibility of the use of green ammonia as a fuel would have a great advantage over other green alternatives such as hydrogen that ammonia is a liquid which can be easily handled and stored at an easily used temperature of $-33°C$. Another advantage of using ammonia is that its combustion does not produce any CO_2. However, any N_2O formed during the combustion process is a more damaging greenhouse gas than CO_2 and it therefore has to be collected and treated. As Haskell points out, there may need to be a radical re-design of the vessels that use it; there will also be added storage and handling problems, ammonia being more toxic than the currently used maritime fuels. However, as shipping currently contributes a large proportion of the world's CO_2 emissions (almost 3% of the global quantity), there is a considerable impetus behind the current movement towards the use of ammonia as a fuel. Haskell reports that it is expected that 7% of the fuel used in ships will be ammonia by 2030 and that this proportion will increase to 20% by 2050.

---

[i] Charles Haskell, Lloyd's Register: https://www.lr.org/en/insights/articles/decarbonising-shipping-ammonia/. This article describes all the advantages of ammonia as a fuel and shows a photograph of an ammonia-fuelled ship.

The introduction of ammonia as a fuel would require very significant increases in global production capacity and the majority of this would need to be produced using renewable resources. Haldor Topsøe has investigated the production of ammonia using hydrogen produced by their SEOC systems. This work has been closely associated with the development of the Syncor reactor discussed in Chapter 4. Hansen and Skyøth-Rasmussen have described a joint project currently being funded by the Danish Energy Agency EUDP in cooperation with Aarhus University, the Technical University of Denmark, Energinet.DK, Vestas Wind Systems, Orsted and Equinor using the SEOC technology described above.[j] They point out that the SEOC-based system has the additional advantage of being able to operate as an energy storage system since the ammonia synthesis reaction can be reversed if necessary. Although the work described in the Hansen and Skyøth-Rasmussen paper is related to the production of ammonia for normal purposes, the product can also be used as a fuel as discussed above. The great advantage of using the SEOC system in conjunction with ammonia synthesis is that the nitrogen needed for the reaction is provided by feeding a mixture of air plus water to the electrolysis system. The product gas is then a mixture of hydrogen and nitrogen; the oxygen of the air feed, as well as that formed by the electrolysis of the water, passes through the solid oxide membrane and therefore also provides a stream of pure oxygen gas. The SEOC therefore in effect also acts as an air separation system. The allover efficiency of this SEOC approach to the synthesis of ammonia is claimed to be ca. 70%.

A press release from the Fraunhofer Institute for Microengineering and Microsystems (IMM) in Mainz, Germany, recently described a project entitled ShipFC aimed at developing eco-friendly technologies for the maritime sector.[k] It describes the development of a catalytic reactor for ammonia splitting that is combined with a fuel cell system for power production. The aim of the team collaborating on the project is to complete a small prototype by the end of 2022 and that a ship powered by an ammonia-powered fuel cell ('The Viking Energy', owned by the Norwegian shipping company Eidesvik) will be put to sea in late 2013.

---

[j] J.B. Hansen and M.S. Skjøth-Rasmussen, 'Can ammonia be a future energy storage solution', Hydrocarbon Processing, January 2020, 1–2.

[k] https://www.fraunhofer.de/press/research-news/2021/march-2021/worlds-first-hightemperature-ammonia-powered-fuel-cell-for-shipping.html.

Green Methanol

As discussed in Chapter 4, methanol is currently synthesised industrially from a mixture of carbon monoxide and hydrogen produced in a well-established industrial process by the steam reforming of natural gas or naphtha. A number of companies are now involved in an alternative process for the production of 'e-methanol' using carbon dioxide and electrolytically produced hydrogen as feedstock. (See Box 4.6.) To date, most e-methanol is produced using electricity from the grid and so the product is not green.

An exciting aspect of the SEOC discussed above compared with the other types of electrolysis systems is that it can also be used to produce a mixture of CO and hydrogen if CO_2 is fed to the system in addition to water:

$$CO_2 + 2e^- \rightarrow CO + O^=$$

The syngas ratio obtained can be adjusted as required by adjusting the H_2O/CO_2 ratio and so the SEOC can in principle be used to provide the feed required for a number of different commercially important reactions such as:

the synthesis of methanol:

$$CO + 2H_2 \rightarrow CH_3OH,$$

methanation:

$$CO + 3H_2 \rightarrow CH_4 + H_2O$$

and the *Fischer Tropsch Process*:

$$nCO + (2n + 1)H_2 \rightarrow C_nH_{2n+2} + nH_2O$$

The consequence of such advances, currently still largely in the development stage, is that it is, at least in principle, possible to produce many important products that are largely currently produced from natural gas by using instead a mixture of water and CO_2 as feedstock. Even though the current costs of such an approach are still relatively high compared with traditional methods, there has been a steady improvement in the performance of suitable SEOC cells so that, if the cost of the renewable electricity used becomes steadily lower (as is currently occurring), the production of totally green products will become commercially attractive. The economic aspects will also become more favourable relatively quickly if CO_2 emissions are taxed at higher levels than at present. Fig. 8.7 shows a schematic representation of the complete technology required, this being based on wind and solar energy, to produce chemicals and 'electrofuels'. The more conventional uses of renewable electrical power are also included for completeness.

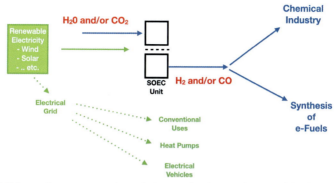

Fig. 8.7 Schematic representation of the production and uses of hydrogen and/or carbon monoxide using an SOEC system.

The 'chemical production' routes shown in Fig. 8.7 include the synthesis of ammonia, methanol and hydrocarbons. For the processes requiring the use of CO such as methanol or Fischer Tropsch synthesis, the CO_2 required as a feed for the SEOC system could be sourced from many different industrial processes if these are available at the required site. However, if the process is to be carried out at a site where such CO_2 is not conveniently available, an alternative route is to capture the required CO_2 from the atmosphere using technology such as that shown in Fig. 8.8. This process has been developed by Climeworks (www.climeworks.com) in collaboration with the company Carbfix (www.carbfix.com). (The latter company is concerned with the permanent storage of CO_2 as carbonate minerals in underground rock structures.) In the example shown, the system is used in conjunction with a geothermal power plant of the type described in Chapter 3 and this supplies the operational energy. The Climeworks process extracts the CO_2 from ambient air by passing it through filters that adsorb it selectively at a temperature around room temperature. When it becomes saturated, the collector is then closed and the temperature is raised to between 80°C and 100°C to desorb the CO_2 (with a purity of 99.7%). This is then cooled to 45°C and collected before delivery for storage or further reaction. Climeworks claims that their filters are fully recyclable and that a single unit can capture approximately 50 t of CO_2 per year directly from the atmosphere.

Use of green hydrogen in steel production

Two major industrial sources of greenhouse gases discussed in Chapter 2 that have traditionally come from the use of coal as fuel are the production of iron

212 Sustainable energy

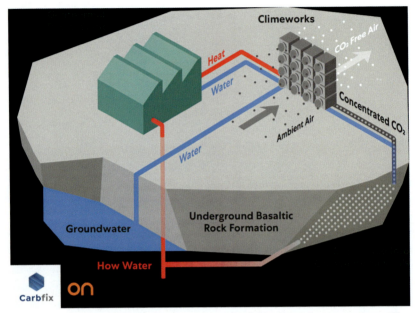

Fig. 8.8 The Carbafix process for the extraction of CO₂ from the atmosphere. *(From (N.d.-b). Climeworks. Reproduced with kind permission of Carbafix and Climeworks.)*

and steel and that of cement. Current progress towards applying green hydrogen as fuel in these processes to minimise the very large emissions of CO_2 that are involved are now discussed.

Fig. 8.9 shows the stages involved in the steel-making process. Iron ore is admitted to the blast furnace together with coke (formed in a coking oven

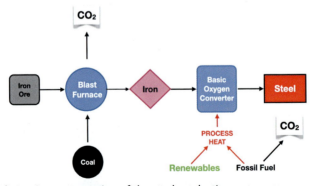

Fig. 8.9 Schematic representation of the steel production process.

prior to admission to the blast furnace) and the latter is converted to carbon monoxide by the residual traces of oxygen present in the feed:

$$2C + O_2 \rightarrow 2CO$$

The CO reduces the iron ore (assumed here to be predominantly Fe_3O_4) to produce pig iron:

$$Fe_3O_4 + 4CO \rightarrow 3Fe + 4CO_2$$

The pig iron is then fed, together with any other required additives (e.g. Cr_2O_3), to the basic oxygen converter where the mixture is converted to steel; the CO formed from the feedstuff is used to provide the process heat as well as a source of the carbon needed for the steel.[l] Scrap iron may also be fed in various proportions at this stage, thus reducing the need for heat energy and therefore cutting the total emissions of CO_2.[m] The emission of CO_2 from the process therefore comes from both the reduction process in the blast furnace and from the combustion of the fossil fuel needed to provide both the process heat and the additional carbon needed in the basic oxygen converter.

It is possible to operate a blast furnace by using natural gas instead of coal and this already gives some improvement in greenhouse gas emissions. Some steel works have already been converted for this purpose. A small number have also been converted to using hydrogen both as a fuel and as a reducing agent and it has been claimed that this already gives 'green steel'. However, as pointed out by the Bellona Foundation (see Box 8.4), the product is only truly green if the hydrogen is produced by renewable electricity and any claims of the production of a 'climate-neutral steel' should be treated cautiously. Most of the plants in Europe using hydrogen for reduction purposes do not appear to be using fully renewable hydrogen.

---

[l] If a solid oxide electrolysis cell (SEOC) such as that shown in Fig. 8.7 were to be used and the feed used for it was water plus CO_2, the system could become largely self-sufficient in carbon.

[m] Scrap iron can be used for the production of steel for many purposes but the product is not pure enough for specialised purposes such as automobile manufacture due to the fact that scrap iron often adds undesirable metal concentrations to the alloy produced. Some 40 years ago, there was a serious problem with the corrosion of automobile bodywork due to the use of recycled steels and this led to the introduction of galvanising by the addition of ZnO to the mix. See for example: https://www.hagerty.com/media/automotive-history/galvanization-sensation-how-automakers-fought-off-the-scourge-of-rust/.

BOX 8.4 Current European use of hydrogen in the steel industry

The Bellona Foundation (https://bellona.org/about-bellona) is a Norwegian organisation with offices in Brussels and also in Russia which aims to provide 'a solution-oriented approach to the environmental challenges'. They have recently published two reports considering the use of hydrogen in steel production in Europe.[n] The first of these considers the use of hydrogen as an auxiliary reducing agent in the 'Blast Furnace - Basic Oxygen Furnace' (BF-BOF) route and the second considers hydrogen as the sole reducing agent in the Direct Reduction of Iron (DRI) route. The first report lists ten companies in Europe who are using hydrogen reduction in BF-BOF plants and it concludes that only three of them are using hydrogen produced by electrolysis while most of the others use grey hydrogen (from natural gas). Bellona queries the claims that these companies are producing green steel, pointing out that using grid electricity in the current situation would actually increase the output of greenhouse gases by 36.7% if a plant was using German electric power. The second report is more hopeful and discusses the situation if iron ore were to be reduced to sponge iron in either a shaft furnace or a fluidised bed reactor using hydrogen produced with renewable electricity. The sponge iron would then be fed to an Electric Arc Furnace (EAF) to which carbon is also fed. Some 14 such plants are planned for Europe in the next ten years or thereabouts and a further ten are in operation. Of the latter, most are most probably currently using natural gas while only two or three are already using hydrogen from electrolysis. Even in the latter cases, it is not yet clear that renewable electricity is being used.

[n]Hydrogen in Steel Production: What is happening in Europe, Parts 1 and 2: https://bellona.org/news/climate-change/2021-03-hydrogen-in-steel-production-what-is-happening-in-europe-part-one and: https://bellona.org/news/industrial-pollution/2021-05-hydrogen-in-steel-production-what-is-happening-in-europe-part-two.

A 2020 briefing prepared for the European Parliament[o] has pointed out that replacing coal with hydrogen would at current prices drive the cost of steel up by one-third. In order to produce the hydrogen by electrolysis would require a 20% increase in the current generation of electricity and hence an even greater increase of renewable production, going beyond the replacement of current fossil-fuel-based electricity generation systems. The briefing also points out that an increased price of steel is likely to cause a shift towards the use of alternative greener materials and that the steel industry would no longer be tied to geographic regions in which coal is plentiful.

[o] https://www.europarl.europa.eu/RegData/etudes/BRIE/2020/641552/EPRS_BRI(2020)641552_EN.pdf.

The situation in Europe regarding a wish to use renewable energy in steel production (Box 8.4) is reflected in many other countries. For example, there has recently been significant interest in reports from Australia that renewable electricity (produced using solar power) will be used to produce hydrogen for steel production by electrolysis.[p] This change would enable Australia's steel production to be concentrated there rather than retaining the current practice of exporting iron ore to China for the production of steel that is then returned to Australia for its domestic use. Not only would such a change allow significant decreases in China's CO_2 emissions but there would also be substantial reductions arising from lower associated maritime emissions during transportation in both directions.

Use of green hydrogen in cement production

Chapter 2 gave a brief description of the process involved in cement production and showed that this industry is responsible for between 8% and 10% of the world's annual emissions of CO_2, a total of almost 900 kg of CO_2 being emitted for every 1000 kg of Portland Cement produced. The main producers are China, India and the US. Coal combustion is generally used to provide the energy needed to bring about the endothermic decomposition of the $CaCO_3$ feedstock, the combustion itself causing significant CO_2 emissions. However, the decomposition reaction itself:

$$CaCO_3 \rightarrow CaO + CO_2$$

is responsible for almost 70% of the CO_2 emission. Hence, even if sustainable energy is used to give green hydrogen that is then burnt to give the necessary energy for the decomposition reaction, only approximately 30% of the total CO_2 emissions will be removed.[q] The use of sustainable energy can also give significant decreases in the substantial quantities of CO_2 emitted in the transportation and handling of materials to and at the cement plant. However, there is no way of cutting down the emissions from the decomposition reaction. Hence, one partial solution is to find other materials to replace at least part of the CaO in the cement needed for construction purposes worldwide by some other constituent. There is currently significant effort aimed at finding alternative materials for that purpose.

---

[p] See, for example: https://www.forbes.com/sites/kensilverstein/2021/01/25/we-could-be-making-steel-from-green-hydrogen-using-less-coal/.

[q] The excess CO_2 could also be converted in a SOEC system of the type described in an earlier section by feeding it with a mixture of steam and the CO_2 emitted from the calciner. The product syngas could then be used for methanol production or Fischer Tropsch synthesis.

However, as there is always likely to be some CaO used in the product, some method of using the emitted CO_2 in a sustainable fashion must be found. Although the CO_2 formed could be trapped for storage underground, the available storage capacities would soon be exhausted if such a method was used for all global cement production. A possible use for the CO_2 would be to use it as a source of syngas by the combined electrolysis of water and CO_2 discussed above, synthesising either methanol or hydrocarbon fuel from the syngas. This could be done either at each cement plant by establishing suitable facilities there (photovoltaic or wind energy generation plus a fuel synthesis plant) or by setting up a suitable pipeline to an existing fuel-synthesis facility. The fuels resulting would not have been made using 'green materials'; however, since the CO_2 used would otherwise have been emitted during the cement preparation process, there would be an all-over reduction in CO_2 emission.

Other areas for energy savings and for the reduction of greenhouse gas emissions

There are many other industrial processes which make use of fossil fuels for the provision of energy. The vast majority of these could instead use hydrogen or a sustainable fuel such as bio-methanol as a source of energy. Although the oil-processing and the petrochemical industries both use a significant amount of self-produced hydrogen (e.g. that formed in cracking processes), improved savings could result in both industries from the use of significantly more renewable energy.

Substantial quantities of electrical energy are used in data-storage facilities and it is clear that many such users are currently installing solar or wind generation systems to provide the majority of their energy needs. Such facilities also most frequently have substantial battery storage facilities.

A great deal of energy is used for both heating and lighting processes in both commercial and domestic situations and considerable effort is being made to achieve significant improvements. Currently, the vast number of users in both of the latter categories still use fossil fuels for heating purposes, these fuels ranging from coal in some areas to kerosine and natural gas in others. All of these energy sources must be phased out and renewable electricity or heat exchange systems such as those described in Chapter 3 must be used instead. There are major moves in many countries to improve the use of energy in commercial and domestic buildings. As an example, Ireland has a very ambitious programme to improve the energy efficiencies of existing housing and has

The way forward: Net Zero **217**

a significant grant programme administered by the Sustainable Energy Authority of Ireland (https://www.seai.ie/grants/home-energy-grants/). Scotland has recently announced an ambitious programme for the installation of heat pumps (https://www.pv-magazine.com/2021/06/14/scotland-announces-massive-plan-for-heat-pump-deployment/) and a similar plan has been announced for the Netherlands (https://www.pv-magazine.com/2021/04/23/massive-plan-for-hybrid-heat-pump-deployment-in-the-netherlands/). The main problem in all of such programmes is that it is difficult to convince existing homeowners that they should upgrade their energy use. However, significant advances are being made in that regard (see Box 8.5) and much more effort of this sort is planned.

Many countries have plans to promote the widespread use of electric vehicles and the use of renewable electricity. However, one of the major problems associated with such vehicles is that of recharging when on an extended journey and, despite the steadily increasing ranges possible on a single charge as discussed in Chapter 6, this is a drawback when it comes to encouraging everyone to change to such vehicles. Further, unless a substantial proportion of the electricity supplied from the grid is renewable, the use of hybrid vehicles gives better figures than those for electric vehicles in many countries. If the subsidiary recharging motors used in hybrid vehicles were powered by biomass-derived fuels, then their range is likely to continue to exceed that of fully electric vehicles and to be significantly

---

BOX 8.5 Irish household energy savings

Ireland has a very significant national programme aimed at encouraging the use of sustainable energy in all regions of the country, both urban and rural. Individual householders are encouraged to upgrade their home energy systems in order to decrease greenhouse gas emissions and, in parallel, to cut their energy costs. Many households have installed photovoltaic (PV) systems and heat exchange systems of the type discussed in Chapter 3 and have also improved their home insulation in order to improve the all-over efficiency of their energy systems. In the case now discussed, the homeowner has recently installed PV panels as well as a heat exchange system that is used for water heating. He also uses a battery vehicle which is now charged using electricity generated with the PV panels when this is available. (His PV system includes battery storage which allows for some load balancing.) Fig. 8.10 shows a recent record of the operation of his PV system that illustrates very clearly how such a system operates effectively even in a country at a relatively northerly latitude (ca. 53°N).

Continued

BOX 8.5 Irish household energy saving—cont'd

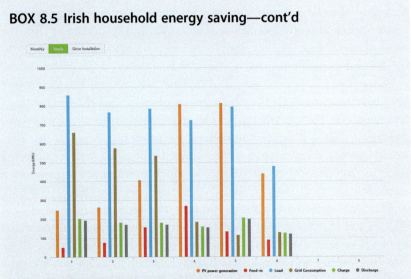

Fig. 8.10 Data for the operation of a domestic Irish photovoltaic system from January to mid-June 2021. *(Data kindly provided by F. Sheehan. Shown are the monthly figures for power generation (orange), feedback to the grid (red), household energy use (blue), draw-down from the grid (brown) and feed to (green) and from (grey) the battery storage system. The maximum household energy use in January 2021 was about 770 kWh.)*

These data, which record the operation of the solar panels on a monthly basis in the period January to mid-June 2021, show that the panels only generate a relatively small proportion of the energy needed for the household (averaging approximately 800 kWh per month) in the early months of the year but that they provide a large proportion of the household load from April onwards. (Data for the period following the commissioning of the system at the end of July 2020 that are not included in this figure show that the PV generation fell off significantly only from November onwards.) The battery storage included in the system has the advantage that when the use of grid electricity is necessary during the winter months, it can mostly be downloaded and stored during the night when the electricity provided is less expensive and also very largely renewable.

more sustainable. Biomass-derived fuels will also play an increasing part in transport, particularly in aviation.

Another area is of major concern: agricultural emissions. Ireland has a particularly significant problem in this regard as it depends very significantly on its agricultural activity, particularly in relation to cattle and dairy products. Recent reports show that Ireland has fallen significantly behind its targets in reducing the emissions from that sector, these contributing 34% of the county's total greenhouse emissions; these agricultural emissions have risen significantly over the last few years rather than falling. Only two other European countries, Hungary and Poland, have shown increases in agricultural emissions over the same period, while Greece, Croatia and Lithuania have shown decreases, these as a result of reductions in their dairy herds. Some methods of decreasing the Irish agricultural emissions are discussed in a very recent report,[r] these including improvements in the efficiency of nitrogen utilisation as well as the use of protected urea products and low emission slurry spreading. However, several other possibilities are not included in that report. For example, it would seem that increasing use should be made of anaerobic digestion methods (see Box 8.6) to reduce the quantities of agricultural residues such as slurry since

---

BOX 8.6 Methanation of biogas

The production of biogas by anaerobic digestion of biological waste was discussed in Chapter 5. The gas from a simple biogas plant contains a mixture of CO, CO_2, H_2 and CH_4 and this can be used directly for heating purposes or can be added in relatively low proportions to natural gas pipelines. However, if green hydrogen can be produced at competitive prices at each plant using the electrolysis methods as described earlier in this chapter (see, for example, Fig. 8.7), it could be used to methanate both the CO and CO_2 components of the biogas mixture to produce high purity methane that could be distributed undiluted in existing gas pipelines as a totally green product.

One of the main sources of greenhouse gas emissions is agriculture and there is therefore a great need to reduce these emissions. An acceleration of the installation of anaerobic digestion systems to provide biogas on farms should therefore be a priority for countries such as Ireland whose agricultural greenhouse gas emissions are currently very high. Ireland has at present just above 30 biogas plants and it has been estimated that it would need to have an additional 900 such plants by 2050. Unfortunately, there are signs of a reluctance to accept such plants by the inhabitants in many areas, it being argued that such plants cause local environmental and safety issues: a NIMBY approach.[s]

[s]NIMBY = 'Not in my back yard'.

---

[r] https://www.epa.ie/publications/monitoring–assessment/assessment/state-of-the-environment/ irelands-environment-2020—an-assessment.php.

220 Sustainable energy

the spreading of this slurry for fertilisation purposes is a very major contributor to greenhouse gas emissions. As the climate in Ireland is particularly suited to the growth of grass, it would be possible for many farmers to concentrate more on the production of silage intended as a feedstock for such anaerobic digestion plants. Further, the report does not address the problem of the steady continued increase in the cattle population in Ireland that has occurred since the removal of milk quotas in 2015. An alternative method of decreasing the agricultural greenhouse emissions, particularly those of biogenic methane, would be to reduce the size of cattle herds and to encourage increased attention to the growth of suitable sustainable energy crops as an alternative to dairy and beef husbandry. As with many of the actions needed in relation to all the sectors discussed in this chapter, decisive government action is needed to attain the necessary reductions in emissions.

Conclusions

The objective of this book is an attempt to give a brief introduction to many aspects of the production and uses of energy as well as an indication of some current approaches to the improvement of energy efficiency in a number of different applications, the aim throughout being to highlight approaches to the reduction of emissions of greenhouse gases. The situation changes from day to day and new reports on governmental and industrial activities in emission reduction are appearing very regularly. What is very clear is that each government must engage fully in the process and must apply a 'carrot and stick' approach to reducing its own emission figures. It is hoped that this book will supply some of the background information needed by those involved in making governmental decisions as well as indicating to academic and industrial researchers some of the areas which require the greatest scientific and technological inputs.[t]

---

[t] It is the author's intention to add in due course short reports on new developments related to some of the topics discussed in this book on his blog at www.contemporarycatalysis.wordpress.com.

Tailpiece

The concept for this book emerged several years ago when I began recognising that various reports outlining plans for the reduction of greenhouse gas emissions failed to include significant details of the chemistry of how these reductions might be achieved. A trivial example was a statement that hydrogen would be produced by reforming natural gas; the report that I read implied that hydrogen production from methane was a straightforward and simple process involving an uncomplicated decomposition of methane. In another item, it was stated that a fleet of city buses in Ireland would be powered by hydrogen produced entirely from indigenous resources; however, no feasible explanation was given as to how that would be achieved, and Ireland does not currently have any appropriate technology. I therefore felt that it would be helpful to write a book that outlines some of the possible methods for controlling greenhouse gas emissions, with the emphasis being placed on the chemistry associated with such methods. My initial intention was that I would first consider only the chemistry behind currently operating processes for the production of energy and then go on to discuss new technologies. My plan was to concentrate on areas in which I had particular knowledge and that were closely related to my own research activities, namely processes involving heterogeneous catalysis. However, I soon realized that I also needed to cover topics about which I have had somewhat less hands-on experience. Hence, the writing process also became a learning experience. I found that I had to expand my prior knowledge of other related topics such as the construction and operation of batteries; the increasing use of photovoltaic, wind, hydroelectric and tidal energy; and even of improvements in domestic heating systems. I hope that at least some part of this rather wide range of content will prove informative to the reader. Further, I hope that specialists in any of the topics that I have covered who feel that my treatment of their subjects is relatively trivial will excuse my attempts at producing easily understandable and readable explanations. Further information on many of the topics handled is easily accessible by following the many links included.

During the final stages of preparation of this book, I exchanged some interactive emails with my friend Prof. Dr. Miguel A. Bañares of CSIC-Instituto de Catalisis, Madrid, Spain, regarding the subjects to be included in my book. During this correspondence, he introduced the analogy of

'Mount Sustainable', representing the huge barrier that must be conquered before the nations of the world can reach "Net Zero" by 2050. This analogy of climbing a mountain is very appropriate: we are not going downhill towards a target; instead we are going to have to surmount a whole series of very significant barriers and even dead ends in our efforts to achieve the summit and the final goal. There are groups of mountains in almost all the countries of the world, some relatively simple to climb, some requiring good guidebooks and some requiring experienced guides, but to climb them all requires dedication and planning. Extending the analogy to climate change, each nation of the world has a number of differing 'Mount Sustainables'. The task faced by each is to overcome a series of their own 'Mount Sustainables', each of these peaks requiring a different approach to the summit. This must be done in a collaborative manner such that each nation ends up by the year 2050 having conquered all of its summits. Teamwork is required for each ascent, especially if all are to be conquered in this relatively short period. The nations possessing less experience will require both financial and technological assistance from those fortunate enough to have much more.

A group of some of Ireland's highest mountains, the MacGillicuddy's Reeks, situated in the Killarney National Park, County Kerry.

Climate change mitigation is on everyone's minds and there are daily reports highlighting one or the other aspect of the subject. The United Nations Climate Change Conference of the Parties (COP 26), being held in Glasgow (31 October–12 November 2021), is about to start. It is to be hoped this gathering will be more than just a talking shop and that agreement may be reached there between the participating nations on the steps that now need urgently to be taken to reach the targets set out in the Paris Agreement of 2015.

Even during the months since I finished writing this book, many documents relevant to its content have appeared, many related to the COP 26 objectives. The latter reports describe in some detail how different nations and industrial groupings hope to achieve the essential reductions in greenhouse gas emissions. It is my intention to attempt to provide updates on some of the recent developments related to the content of the book on my blog at https://contemporarycatalysis.wordpress.com/. I encourage you-the reader-to register and login to this blog and even to consider contributing additional items for inclusion.

Julian R.H. Ross, October 2021

Postscript

During the final stages of publication of this book, it has proved possible for me to write a brief additional note on the outcome of the COP 26 gathering. This meeting, held in Glasgow at the beginning of November 2021, was attended by the leaders of most of the nations of the world and received much attention from the world press. Towards the end of the meeting, it appeared that full consensus had been reached regarding all the items for which a unanimous decision had been required. However, in the final moments of the closing session, three countries—Australia, China and India—withdrew their support for a decision to cease the use of coal by 2030. (Following the mountaineering analogy of the Postscript, an insurmountable ravine had been encountered just below the summit of 'Coal Mountain'.) Nevertheless, some very important decisions were reached during the meeting, not only within the conference walls but also through parallel international negotiations, regarding how the objectives of the Paris Agreement might be reached. It was agreed in Glasgow that most of the Paris objectives should be significantly accelerated between now and 2030. All the participating nations have agreed to submit updated plans as to how these new objectives will be achieved in their own countries. (A good summary of the main conclusions of COP 26 is to be found at https://ec.europa.eu/commission/presscorner/detail/en/ip_21_6021.)

It is abundantly clear that now is the time for action on all fronts and that we must all now move more firmly from words to action. It is also clear that such action will be costly to all involved. Governments must help and encourage all their citizens and corporations to work towards the reduction of greenhouse gas emissions on all fronts and must impose sanctions when necessary on those who do not contribute to the required reductions. Many of the methods of reducing greenhouse gas emissions described in this book are well advanced and are immediately applicable when appropriate, but some are still under development. Much supporting research and development in the field of energy will therefore be required over the next decade.

Julian R.H. Ross, November 2021

Index

Note: f = figure; t = table; b = box

A

Agricultural greenhouse gas emissions, reductions in, 218–220
Algae, 126–127
Ammonia production, 99–102, 100–101f, 205–220
An approach to Net-Zero
  electrolysis, 200b
  fuel cells, 203–204
  national reports on, 202–203b
  renewable energy, 199–203
Autoreforming, 91–93, 92f

B

Battery electrical vehicles, 154–159, 156–157f
Battery-operated vehicles, 133–134
Biodiesel, 103–104, 117–118
Bioethanol production, 113–117, 114t, 115–116f, 117t
Biogas, methanation of, 219b
Biomass
  algae, 126–127
  chitin, 127–128, 128f
  gasification, 123
  microalgae, 127, 128f
  non-traditional uses
    ethanol/bioethanol production, 113–117, 114t, 115–116f, 117t
    fuels and chemicals, 119–122, 120f, 122f
    lignocellulosic crops, 119–122, 120f, 122f
    oil-based residues, 117–118, 118f
    oil crops, 117–118, 118f
    organic grasses, 112–113
    organic residues, 112–113, 113f
    pyrolysis, 123–125, 124–126f
    seaweed, 126–127
    soil organic matter, 129b
    wood, 104–111, 106b

  paper production, 107–109, 107t, 108f, 110f, 110b
  paper recycling, 110–111, 111b
Blue hydrogen, 77, 78b, 80b, 94, 199

C

Carbon capture and storage (CCS), 24, 80b, 96
Carbon dioxide, 6, 7b, 7f
Carnot cycle, 141b
Cement production, 33–34, 33b
Chemical production routes, 211
Chitin, 127–128, 128f
Chlorofluorocarbons, 9
Coal
  cement production, 33–34, 33b
  electricity generation, 26–33, 27f, 29b, 29t, 30–31f, 32t
  heating combustion, 23–26, 24–25f
  iron production, 34–38, 35t
  power generation, 26, 26f
  steam engine, 26, 26f
  steel production, 34–38, 35t, 37b
Colour-coded nomenclatures, 78b
Crude oil, 38–41, 39–40t, 42f

D

Daniel Cell, 175, 175f
Diesel engine, 136–137, 137f, 141, 142f, 143b
Dry cell batteries, 178, 178f

E

Electrical vehicles, 131
Electric arc furnace (EAF), 214b
Electricity generation, 26–33, 27f, 29b, 29t, 30–31f, 32t
Electrochemical batteries
  Daniel Cell, 175, 175f
  dry cell batteries, 178, 178f

225

226 Index

Electrochemical batteries *(Continued)*
  lead acid battery, 176–177
  Li-ion batteries, 181
  lithium metal batteries, 184
  rechargeable nickel-cadmium (Ni-Cd) batteries, 179
  rechargeable nickel-metal hydride (Ni-MH) batteries, 179
Electrochemical processes, kinetics of, 170–174
Electrochemical series, 163–170
Electrolysis, solid oxide fuel cell (SOFC), 192–195, 193–194f
Ethanol production, 113–117, 114t, 115–116f, 117t
Exchange current density, 170–171, 171–174b
Exhaust emission control, 143–150, 144f, 146f, 148f, 151f, 151b
European emission statistics, 12–15, 13t, 14f, 15t, 17–18, 17f
European Environment Agency, 11–12

F
Faraday Laws, 163–166
First-generation biorefineries, 112–128
Fischer Tropsch process, 97–99
Flow batteries, 185–186, 185f
Fuel cells, 186–192, 186f, 188t
  advantages, 205b
  methanol fuel cell vehicle, 205f
  molten carbonate fuel cell (MCFC), 190, 190f
  polymer electrolyte membrane fuel cell (PEM), 187, 189f
  solid oxide fuel cell (SOFC), 191–192, 191f
  for transportation, 203–204
  use in transportation vehicles, 159–160, 160f, 203–204
Fuel-cell vehicles, 159–160

G
Gasification, 123
Geothermal energy, 55–59, 56–57f, 57t
Global warming, 1–2, 2f
Green ammonia, 206–209

Greenhouse effect, 3, 3f, 9–11, 12f
Greenhouse gases, 4–9, 4–5t
  carbon dioxide, 6, 7b, 7f
  chlorofluorocarbons, 9
  methane, 6–8, 8b
  nitrous oxide, 8
  ozone, 9
  sources, 11–19, 13t, 14f, 15t, 16–18f
  water vapour, 5–6
Green hydrogen, 77
  cement production, 215–216
  steel production, 211–215
Green methanol, 210–211
Grey hydrogen, 77

H
Half-cell EMF, 167–170
Heating combustion, 23–26, 24–25f
Household energy savings, 217–218b
Hybrid vehicles, 150–152
Hydroelectric power, 63–67, 64–65b, 65t, 66f
Hydrogen
  ammonia production, 99–102, 100–101f
  autoreforming, 91–93, 92f
  from biomass, 96
  blue hydrogen, 77, 78b, 80b, 94, 199
  carbon capture and storage (CCS), 80b
  conventional preparation, 88–89
  economy, 77
  endothermic reversible reaction, 79–81
  Fischer Tropsch process, 97–99
  fuels production, 97–99
  green hydrogen, 77
  grey hydrogen, 77
  membrane systems, 89, 91b
  methane
    dry reforming of, 93
    partial oxidation of, 89, 91–93, 92f, 99
    pyrolysis of, 93–94
    steam reforming of, 77–89, 78b, 79f
  methanol production, 96–97, 98b
  microreactors, 86b
  nomenclatures for H$_2$, 78b
  pressure-swing absorption (PSA) system, 81
  production costs, 77, 96

production using renewable energy, 199–203

production by water electrolysis, 94–95, 95*f*

sophisticated high-pressure transporters, 81–84

I

Intergovernmental Panel on Climate Change (IPCC), 9–10

Internal combustion engines, vehicles with, 134, 135–136*f*, 138

Iron production, 34–38, 35*t*

K

Kyoto Protocol, 9–11

L

Lanthanum-strontium-cobaltite, 206

Lanthanum-strontium-ferrite-cobaltite, 206

Lead acid battery, 176–177

Lignocellulosic crops, fuels and chemicals from, 119–122, 120*f*, 122*f*

Li-ion batteries, 181

Lithium metal batteries, 184

M

Membrane systems, 89, 91*b*

Methanation of biogas, 219*b*

Methane, 6–8, 8*b*

dry reforming of, 93

partial oxidation of, 89, 91–93, 92*f*, 99

pyrolysis of, 93–94

steam reforming of, 77–89, 78*b*, 79*f*

Methanol production, 96–97, 98*b*, 205–220

Microalgae, 127, 128*f*

Microreactors, 86*b*

Molten carbonate fuel cells (MCFCs), 190, 190*f*

N

Natural gas, 42–47, 43–44*t*, 45–46*f*, 47*t*, 89–96

dry reforming of, 93

partial oxidation of, 89, 91–93, 92*f*, 99

pyrolysis of, 93–94

steam reforming of, 77–89, 78*b*, 79*f*

Net Zero by 2050, 197

Nitrous oxide, 8

Nobel Prize, Chemistry, 182–183*b*

Nuclear energy, 50–55

cold fusion, 55*b*

nuclear fission, 51, 52*f*

nuclear fusion, 51–53, 52*f*

nuclear power, 53–54, 53*f*

radioactive decay, 51

O

Oil-based residues, 117–118, 118*f*

Oil crops, 117–118, 118*f*

Organic grasses, 112–113

Organic residues, 112–113, 113*f*

Otto engine, 139, 139–140*f*, 141*b*

Overpotential, 170–171, 171–174*b*

Ozone, 9

P

Paper

production, 107–109, 107*t*, 108*f*, 110*f*, 110*b*

recycling, 110–111, 111*b*

Paris Accord, 11, 12*f*, 197

Paris Agreement, 9–11

Petroleum, 41, 42*f*

Photoemission, 69–70

Photovoltaic generation of electricity, 69–70, 72

Plug-in hybrid vehicles, 152–154, 153*f*

Polymer electrolyte membrane fuel cells (PEMs), 187, 189*f*

Power generation, 26, 26*f*

Pressure-swing absorption (PSA) system, 81

Pyrolysis, 123–125, 124–126*f*

R

Rechargeable nickel-cadmium (Ni-Cd) batteries, 179

Rechargeable nickel-metal hydride (Ni-MH) batteries, 179

Renewable energy

geothermal energy, 55–59, 56–57*f*, 57*t*

hydroelectric power, 63–67, 64–65*b*, 65*t*, 66*f*

nuclear energy, 50–55

solar power, 69–74, 70–74*f*

228 Index

Renewable energy *(Continued)*
  tidal energy, 59–62, 59–61*f*
  wave power, 62–63, 63*f*
  wind power, 67–69, 67–68*f*, 68*t*

S

Salt bridge, 170*b*
Seaweed, 126–127
Second-generation biorefineries, 112–128
Soil organic matter, 129*b*
Solar activity, 1–2, 2*f*
Solar power, 69–74, 70–74*f*
Solid oxide fuel cells (SOFCs), 191–195,
    191*f*, 193–194*f*, 206
Solid oxide hydrolysis cells
  energy savings, 216–220
  greenhouse gas emissions, 216–220
  green hydrogen
    ammonia, 206–209, 207*f*, 214*b*
    cement production, 215–216
    steel production, 211–215, 212*f*,
        217–218*b*
  methanol, 210–211, 211–212*f*
Sophisticated high-pressure transporters,
    81–84
Standard electrode potentials, 168*t*
Steam engine, 26, 26*f*
Steel production, 34–38, 35*t*, 37*b*

T

Tidal energy, 59–62, 59–61*f*
Transport
  battery electrical vehicles, 154–159,
    156–157*f*

diesel engine, 136–137, 137*f*, 141, 142*f*,
    143*b*
electrical vehicles, 131
exhaust emission control, 143–150, 144*f*,
    146*f*, 148*f*, 151*f*, 151*b*
fuel cell vehicles, 159–160, 160*f*
historical development, 131–143
hybrid vehicles, 150–152
internal combustion engines, 134,
    135–136*f*, 138
Otto engine, 139, 139–140*f*, 141*b*
plug-in hybrid vehicles, 152–154, 153*f*

U

United Nations Climate Change
    Conference of the Parties (COP 26),
    Glasgow, 222–223
United Nations Convention on Climate
    Change (UNFCCC), 9–10, 197

V

Volta pile, 163–166, 164*f*

W

Water electrolysis, 94–95, 95*f*
Water vapour as greenhouse gas, 5–6
Wave power, 62–63, 63*f*
Wind power, 67–69, 67–68*f*, 68*t*
Wood, 104–111, 106*b*
  paper production, 107–109, 107*t*, 108*f*,
    110*f*, 110*b*
  paper recycling, 110–111, 111*b*

Printed in the United States
by Baker & Taylor Publisher Services